RIVER DON

Elizabeth Reeve

AMBERLEY

First published 2015

Amberley Publishing
The Hill, Stroud, Gloucestershire, GL5 4EP
www.amberley-books.com

ISBN 978 1 4456 3868 3 (print)
ISBN 978 1 4456 3885 0 (ebook)

British Library Cataloguing in Publication Data.
A catalogue record for this book is available from the British Library.

Typesetting by Amberley Publishing.
Printed in the UK.

CONTENTS

The Don catchment.

INTRODUCTION

Rivers are the lifeblood of the world, collecting and storing rainwater by which all life on earth is able to exist. They provide breeding grounds for both freshwater and marine organisms, many of which in turn provide food for human beings.

Throughout the ages, they have enabled navigation and, since the Industrial Revolution, power to drive machinery. In our turn, our most creative minds have invented structures and engines to take advantage of this power, often destroying the very water of life in the process.

The River Don is one such. With its many tributaries feeding its 70-mile length on its journey to the sea, it has undergone many changes over its lifetime. From a sparkling brook to what has been called a sewer, the Don has suffered greatly. For thousands of years, it meandered from the heights above Penistone in all its natural beauty, until the nineteenth century when man recognised its power and used its growing strength for his own commercial gain, giving little thought to the damage that was being done.

By the 1950s, much of this damage was evident. Fishing was no longer possible or advisable, as the pollution related to Sheffield's steel industry, among others, had turned that sparkling stream into a filthy, foam-filled cesspit. Weirs and dams had been built to control its flow and mills and mines littered its length, pouring contaminants of all sorts into its waters.

By the 1960s and '70s, action was beginning to be taken to turn the tide, so that fifty years on we are again seeing salmon and many species of freshwater fish fighting their way upstream to their spawning grounds. Fish and eel passes are being installed to help them up the weirs and small mammals are returning.

As someone who was brought up by the riverside, saw the devastation that industry caused and returned in retirement to see the changes that have been wrought in the interval, I can only welcome the work of the various river authorities that have been involved in this reincarnation. They might not do everything right all of the time, but the benefits to the River Don show it is better to do something than nothing at all.

Liz Reeve
17 February 2014

THE RIVER DON

The title of this book is something of a misnomer, as the River Don doesn't actually reach the sea in its own right; it finally empties into the River Ouse at Goole before travelling onward to the Humber estuary and thence the sea. It is also the receptacle of many rivers and smaller tributaries that feed into it as it flows in a south-easterly direction from the Pennines towards Sheffield, all of which have had an impact on the Don in one way or another. The Little Don, Ewden Beck and the Loxley all join the Don before it reaches Sheffield, and the River Sheaf emerges in the city itself.

The River Rother, rising in the Peak District National Park to the south, and the Dearne, from the west near Huddersfield, both reach the Don before it arrives in Doncaster. The main tributaries thereafter are the Ea Beck and the River Went, both feeding into it from the west as it flows in a northerly direction to the Ouse.

It can be seen, therefore, that the River Don plays a very important part in the drainage of a substantial part of England. As the rainy month of June 2007 proved, when approximately 30,000 homes and businesses in South and East Yorkshire had to be evacuated, putting many thousands of people out of their homes for months, an excess of water draining into the Don can cause flooding on a massive scale. It behoves the Environment Agency to ensure that measures are in place to enable it to cope should a similar situation arise in the future.

There are three distinct parts to the River Don, each of which is used for a specific purpose. The upper reaches are devoted to the public water supply, with reservoirs used as storage; the middle section, in which weirs were built for the powering of machinery; and the final section, which is used for navigation, locks being constructed to deal with variations in water levels.

The whole of this area has now been brought together under the co-ordination of the Don Catchment Rivers Trust, which is endeavouring by various means to rectify the damage caused over hundreds of years by industry and the influx of population required to service this. Sheffield was the main centre of this explosion during the Industrial Revolution in the mid-1800s, but since its demise as the centre of the steel industry and the implementation of the Rivers Prevention of Pollution Acts of 1951 and 1961, and the Clean Air Act in the 1960s, great strides have been made to overturn the damage done there, which made the River Don one of the most polluted in Europe, and Sheffield probably one of the dirtiest cities.

Until the 1700s, navigation on the Don was very limited and attempts to improve it were opposed for many years, but the instigators won out in the end and navigation is now possible from Sheffield to Goole following the building of canals. Unfortunately, the final upgrading of the canals and locks, which began in the 1980s, came too late as river trade collapsed and the canals are now mainly used for leisure.

In the seventeenth and eighteenth centuries, salmon were so prevalent in the Don that they were everyday food for even the poorest, but they declined steadily until they disappeared altogether. However, with improvements in water quality and wildlife habitat since the 1950s, and with the more recent installation of fish and eel passes, all fish species are gradually returning, to the extent that angling clubs flourish and the possibility of salmon in the Don is no longer a forlorn hope.

The River Don at Lower Sprotborough.

1

A PUBLIC WATER SUPPLY

THE SOURCE

As might be imagined, finding the source of a river in such inhospitable moorland as can be found above Penistone in the northern Peak District was never going to be easy, even on such a beautiful day as it was when my husband and I set out to do just that in May 2014.

A visit to Penistone Library unearthed an updated version of *The Tour of the Don Vol. 1*, written by John Holland in 1837. It describes his search for the Don Well, which was reputed to be the source, which he eventually found on the southern side of Snailsden:

> On the southerly side of the mountain, we have what is called Dearden Water and running into which, at some distance above its union with the Swinehul, are two gullies called grains (a Scottish name for small divergent water courses) … It is upon the higher of these grains that we find the celebrated Don Well.

The well is described as,

> a hole in the bank side, apparently about 20 inches in diameter, and a dozen yards from the channel of the stream with which its beautifully pellucid water quietly mingles, after flowing down a slope deeply matted with long grass and the well-known bog-moss.

However, while this spring was considered to be the source of the River Don, it became apparent to me that rainwater would also drain from the surrounding moors to swell the stream, and that Snailsden Reservoir was ideally situated to receive it. There are in fact three reservoirs through which the Don passes at the beginning of its journey, and our search began with Winscar, the nearest to Penistone, but the third in line from the source.

Even finding Winscar was not as easy as we had thought it would be – signage being limited – and we found ourselves in Holmfirth before we realised we had passed the turning. However, our perseverance was rewarded, as, on alighting from the car, we were delighted to be met by the first cuckoo call we had heard for several years. Walking to the dam wall, we discovered our first view of the beginning of the river, as it flowed from the overflow channel of the reservoir to the valley below and from where it continued in an easterly direction towards Penistone.

On our way back to the car we were entertained by another visitor to the reservoir, who reminisced about visiting the nearby Tinker's Monument in his childhood. With a treacle sandwich in his pocket and a penny to go up the tower, he and his friends would be sent off by their mothers for a day out in the country; the freedom of children in the past bears no resemblance to that allowed today. Leaving him to enjoy the view, we followed his directions to Snailsden, and as we climbed back to the main road, we could just see the second reservoir, Harden, away to our left.

A few miles further on, travelling on a narrow country road towards the highest part of the moors, we caught a glimpse of a noticeboard announcing that we had reached Snailsden Reservoir. All the reservoirs are managed by Yorkshire Water and posters warn of the dangers of cold water; swimming is definitely not allowed. Turning around and retracing our route as soon as possible, we managed to park in the gateway before climbing over the stile built into the wall and walking up a path lined with beautiful dry-stone walls and bilberry plants.

Reaching the top, we were met with the sight of the reservoir surrounded by hills and, as at Winscar, the overflow channel running down into the valley below where the stream could just be seen in the bottom heading towards Harden and Winscar.

Winscar Reservoir.

Harden Reservoir.

On such a lovely sunny day, it was a pleasure to be there and we were able to admire the stark beauty of the moorland, while not underestimating the dangers that might prevail in less favourable weather conditions.

DUNFORD BRIDGE

Returning past Winscar, we left the reservoirs behind and drove into the valley towards Dunford Bridge – the village we had looked at over the dam. On closer inspection, the place we arrived

Left: Snailsden stile.

Below left: Snailsden Reservoir.

Below right: View from Winscar.

at did not quite live up to the expectations offered by the view from above, but nevertheless it did have other attractions for us.

The car park is an access point to the Trans Pennine Trail, a 250-mile long stretch connecting the North and Irish Seas from Hornsea to Southport, and a good place to park, walk and cycle towards Penistone, from where several directions can then be taken.

The river here is more of a stream than a river, its flow being controlled to some extent by the amount of water in the reservoirs. From the bridge, however, one can see that rather than there being a weir, the water flows down over a series of steps. It was interesting to discover, therefore, that construction of the railway caused the upstream channel to be straightened to facilitate the laying of the railway track, demonstrating that the River Don was suffering physical damage right from its source.

Trans Pennine Trail at Dunford Bridge.

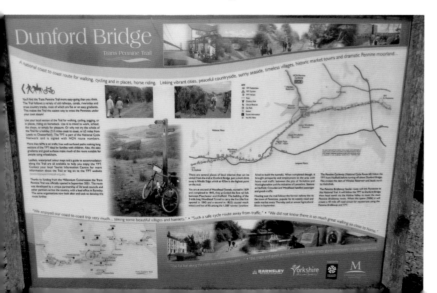

Dunford Bridge steps and forget-me-nots on the riverbank.

With trees on either side of it and lots of greenery in the water, it seemed in danger of becoming choked up here, but the bank was covered with forget-me-nots, which provided a beautiful blue haze. The view from the other side of the bridge was somewhat less attractive, being dark and overgrown.

From the car park it was apparent that work was being undertaken by the Electricity Alliance, which now owns some tunnels, with a noticeboard explaining that cables were being re-routed through them.

For anyone who enjoys a mystery thriller, the novel *Blind to the Bones* by Stephen Booth is set in this remote part of the county and provides additional atmospheric information about the area.

Unbeknown to us until this point, we discovered from a workman that these were the Woodhead Tunnels – famous when we were children as being the longest railway tunnel through which one would pass, in total darkness if the lights failed to come on, in the steam train that took us to Manchester Belle Vue Zoo on Sunday School outings. It was electrified in 1953 and closed in 1981, much to the chagrin of collectors of railway memorabilia.

The original single track, the tunnel for which is currently being blocked up, has now been taken up and provides the route of the Trans Pennine Trail.

Woodhead cable tunnel.

PENISTONE AND OXSPRING

Penistone is renowned as the highest market town in Yorkshire, and caters for every need, be it residential or tourist. However, one should be aware that, befitting its reputation, that description carries an inherent warning of wind, rain and snow and one should not take that too lightly when visiting the moors, even in summer. Despite this, it is an ideal centre for exploring the area.

The river at Penistone is still a relatively narrow and shallow watercourse, but one should never assume that this is always the case. Heavy rain can quickly swell the stream to a raging torrent, and it was for this reason that the weir at Penistone was removed in the 1970s, when it was thought that it contributed to flooding.

Despite being so near to the source of the River Don, pollution from ochre was already in evidence due to the mining of ganister near Penistone. Ganister is a particular kind of sandstone which is used to line furnaces, and leakage of ochre, which was also found in the mine, caused the river to turn orange. A similar situation arose at Beeley Woods north of Sheffield.

Just down river from Penistone is Oxspring Viaduct, which is accessible by a public footpath from the B6462 travelling towards Wortley. Parking in a nearby lay-by and walking down to the river, we arrived at a tree-lined glade where a couple of dogs were enjoying playing in the water, before crossing over a narrow pack-horse bridge to the other side.

Passing under the viaduct, a pleasant walk brought us to the yard of Oxspring Wire Mills, from where the company's driveway brought us back to the road half an hour or so later. The River Don was a source of power here as far back as the thirteenth century, when the lord of the manor's fulling or walk mill was used to shrink woven woollen cloth before it was made into clothes.

OUGHTIBRIDGE

Seven and a half miles from Oxspring, this is another village which has found itself at the mercy of the River Don as it falls towards Sheffield, 5 miles away. With ten more reservoirs feeding into the Don between Winscar and Oughtibridge, it seems inevitable that unfavourable climatic conditions will arise from time to time with damaging effects, and this was the case

on an epic scale on at least two occasions within living memory – in 1947 and 2007.

More Hall Reservoir, the nearest one to Oughtibridge, is the home of a fly fishing club and a circular walk is possible. The 'ladder', over which water runs from the reservoir to the river below, was dry on our visit, but it is long and steep and must be quite a sight when water is released.

There are now a lot less pubs and shops in Oughtibridge than there used to be, but places of interest include Oughtibridge Hall – thought to be the oldest house in the village and dating from 1370 – and the hamlet of Onesacre half a mile out of the village,

Oxspring viaduct.

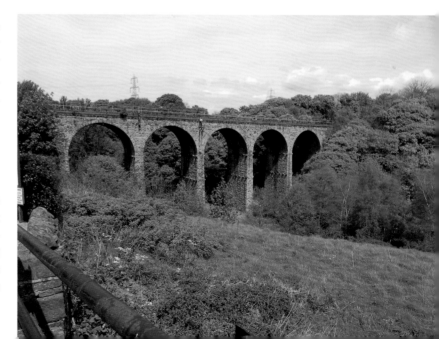

A pack horse bridge and a leafy glade.

which was mentioned in the Domesday Book as belonging to a Saxon lord named Godric. Onesacre Hall, a Grade II listed building, dates back to 1580 and was built by Nicholas Stead and his son Thomas, whose ancestors had acquired the estate in the fourteenth century.

The land on which Coronation Park is located was gifted by a local benefactor in 1911, and created to celebrate the coronation of George V. However, it would appear that it was renamed at the time of the coronation of Queen Elizabeth in 1953, and is looked after by Bradfield Parish Council. Situated on the side of the river, it remains vulnerable to flooding, but between times it is a valuable resource for the community and a group of volunteers work with the council to improve and maintain this attractive green space.

On our visit to the area, the Station Road Bridge, close to the park and the Cock Inn (which boasts pub food and a tea room) was under repair, having been weakened by the 2007 flood. Consequently, the first weir, which is immediately on the down-hill side east of this road bridge, was difficult to access. However, the work will soon be completed and there is a public footpath and a pleasant walk to be had if travellers have time on their hands.

More Hall Reservoir and ladder.

Above: The Don Bridge.

Below: Coronation Park.

2

AN INDUSTRIAL HERITAGE

SHEFFIELD

The river at Oughtibridge is still very shallow, but as it falls 160 feet between Oughtibridge and Brightside in Sheffield, one can well understand the force produced in such circumstances. As a result, industrialists took advantage of this and many mills for flour, paper and snuff proliferated along its route. Weirs were built and water wheels installed for the production of steel, cutlery, silverware, nails and scythes and blast furnaces. Forges, rolling and slitting mills and high pressure steam engines were also operated by water power, some still being used in 1900.

It is hard to imagine what it must have been like as engines pounded, steam belched forth and workers toiled year in and year out to make a living for their families and wealth for their masters.

The dark satanic mills of the hymn 'Jerusalem' were never more in evidence than here. In the 1950s, a pall of smoke and grime covered the city and a bus journey into it from the countryside beyond was like taking a ticket to hell. This may sound over-imaginative, but the blackness of buildings and the smoke and soot emitted, with red and orange flames glimpsed through doors and windows were enough to turn any child's thoughts in that direction.

The aforementioned mills caused part of this, but the growth of steel-making plants throughout Sheffield and into Rotherham, as is demonstrated at Magna, the Science Adventure Centre which is built on the Steel Peach & Tozer plant at Rotherham. Doncaster, too, had its own problems with coal mines being sunk, but Sheffield was the major perpetrator, despite the wealth it yielded.

One of the anomalies of this, however, was a bright spot in the devastating gloom, as it was found that fig trees grew on the river bank in Sheffield due to the temperature of the water rising to 20 degrees following the quenching of hot metals. These trees are now fifty years old, but, following the demise of the steel industry, are not expected to survive indefinitely as no new seeds have germinated since the temperature of the water has dropped.

What is now known as the Five Weirs Walk, starting at Blonk Street, was part of this route and, though most of the mill buildings have now disappeared, the weirs remain as a reminder of former times. While man-made, they nonetheless have a beauty of their own within the dappled light of overhanging trees.

Sheffield's industrial heritage is now managed by the Sheffield Industrial Museums Trust which runs three sites: Abbeydale Industrial Hamlet, Shepherd Wheel Workshop, and Kelham Island Museum. Special events are held throughout the year.

ABBEYDALE INDUSTRIAL HAMLET

Though based on the River Sheaf, which flows in a northerly direction to join the River Don near Blonk Street Bridge in the city, Abbeydale Works was the largest water-powered industrial site on the river, and acts as an example of the type of works which sprang up all along the River Don.

Agricultural tools, such as scythes and sickles, were produced here and exported all over the world.

The hamlet now comprises workers' cottages, water wheels, workshops, a grinding hull and the last complete surviving crucible steel furnace in the UK in Grade I and II listed buildings.

A man-made goit takes water from the Sheaf and diverts it into a dam next to the hamlet, to then be used to turn the water wheels that powered the tilt hammers, grindstones and machinery required to make tools. It then runs back into the Sheaf. It is located at Abbeydale Road South, Sheffield S7 2QW.

SHEPHERD WHEEL WORKSHOP

The Porter Brook, one of the main tributaries of the River Sheaf, is the location of the restored Shepherd Wheel, a unique working example of a water-powered grinding hull. There has been a water-powered workshop here since the fifteenth century but the current workshop buildings date to the eighteenth and nineteenth centuries, and show how the water wheel powered the grindstones. Housed in another Grade II listed building, it is also a scheduled ancient monument.

It is located at Whiteley Woods, Off Hangingwater Road, Sheffield.

Abbeydale hamlet.

Above: The dam and one of the wheels it powers, courtesy of Sheffield Industrial Museum Trust.

Left: One of the Abbeydale water wheels.

KELHAM ISLAND MUSEUM

This museum is situated on an area of land between the River Don and the goit or millstream. The weir dates back to the early thirteenth century, when it diverted water into the millstream for a corn mill. It is named Kelham Island after Kellam Homer, the town's armourer, who had a grinding wheel there in 1637.

The museum tells the story of steel making and how it forged the Sheffield of today, and its impact on the modern world. Its major attraction is the 12,000 hp River Don Engine, the most powerful steam engine in Europe. This engine was built in 1905 and originally ran at Cammell's works, where it rolled armour plate for fifty years before being transferred to British Steel Corporation's River Don Works to continue its working life until 1978. It has now found a home at Kelham Island Museum, where it can be seen 'in steam', though visitors should check before visiting if they wish to experience this.

This is not the museum's only point of interest, however, as it holds many other artefacts relevant to Sheffield's story, including the Hawley Gallery, which has a fascinating display of tools used in the silver smithing and cutlery industry. A 'street' of workshops has also been constructed in recognition of the work of the 'Little Mesters' – self-employed craftsmen who specialised in the individual trades of forging, grinding and finishing tools and cutlery by hand.

It is located at Alma Street, Sheffield S3 8RY. For further information on all these venues, you should contact Kelham Island Museum, Alma Street, Sheffield S3 8RY. Tel: 0114 272 2106. Email: ask@simt.co.uk.

The goit or millstream runs on one side of the museum and the River Don on the other, creating the area known as Kelham Island.

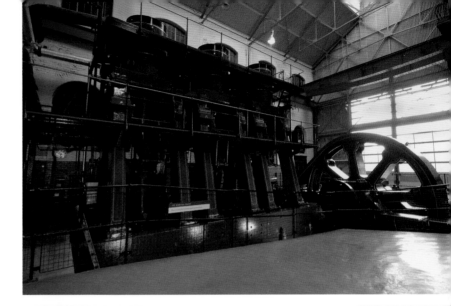

Above: The Don Engine, courtesy of Sheffield Industrial Museum Trust.

Below: New apartments built alongside the River Don at Kelham Island.

Above left: Lady's Bridge in the distance, the Sheaf emerges from a culvert.

Above right: Tinsley Viaduct.

Below: Meadowhall Weir with the shopping centre on the left.

BRIDGES THROUGH SHEFFIELD

Many bridges cross the Don as the river weaves its way between Hillsborough, where a footbridge provides access to the Sheffield Wednesday football ground, and Tinsley Viaduct near the Meadowhall Shopping Centre.

Lady's Bridge, with its five arches, is the oldest bridge crossing the Don in Sheffield, being built in 1485; it now connects the city centre with The Wicker.

The Wicker Viaduct, at 660 yards long, was built in 1845 to carry the Great Central Railway to Manchester across the canal, roads and the River Don. It has twenty-seven arches to the south side and twelve to the north, and bears the weight of Sheffield Victoria railway station, which was built on them.

The origin of the name Wicker is uncertain, though in a tradition dating back to the thirteenth century it was a place where sports and athletic activities were practised.

Strung from The Wicker Viaduct by a web of steel cables is the Cobweb Bridge. At 330-feet long, it saves a detour of a mile as part of the Five Weirs Walk.

Another name to conjure with is the Cadbury's Works Bridge, so called after the famous chocolate maker, though the factory was originally the home of Bassetts Liquorice Allsorts, Jelly Babies and Wine Gums. It was sold again in 2010 when Cadbury was taken over by Kraft.

Tinsley Viaduct is 0.6 miles long and cost £6 million to build in 1968. It was strengthened in 1983 and again in 2006, at a

Above: Blonk Street Bridge.

Below: Millennium Square and the Peace Gardens.

Above left: Rotherham Weir and Tesco.

Above right: The water feature outside Pond Street railway station.

cost of £81 million, when it was reduced to two lanes in each direction following the introduction of an EU directive allowing 40 ton lorries.

I was particularly intrigued by the name Blonk Bridge, which is situated near the Victoria Quays (featured in the next section) and the beginning of the Five Weirs Walk, and discovered it was named after Benjamin Blonk, who, in November 1779, was recorded in the Silver Register of the Sheffield Assay Office as the maker of scissor bows and saddle nails. The culverted River Sheaf also joins the River Don at this bridge. Mr Blonk was one of a group of men who were given fishing rights by the Duke of Norfolk in 1786, for the price of a dish of fish once a year.

The city has been transformed in recent years and is now a very pleasant place to visit.

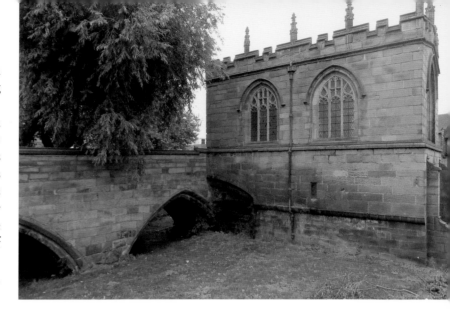

Above: Chantry Chapel of Our Lady on the old Don Bridge.

Below: Children frolic in a sandpit.

ROTHERHAM

Rotherham, 6 miles downstream from Sheffield, sits on the confluence of the Rivers Don and Rother and forms part of the Sheffield Urban Area according to the National Office for Statistics. It is a conurbation with a population of 685,368 (2011 census), which is over half that of the county of Yorkshire as a whole. It has had its share of industry in the past, coal and iron being the major contributors.

A Tesco store now sits between the river and the canal, but its car park was leased by the Walker brothers for their water-powered forge in 1754, since when it has been known as Forge Island. The Rotherham Forge and Rolling Mills eventually became the River Don Stamping Co. Ltd, which ceased to operate in 1981.

The Chantry Chapel of Our Lady, dating from 1483, is a well-known sight on the side of the road in Rotherham. Standing as it did in the middle of the medieval bridge, which was the main road into Rotherham, it provided travellers with an opportunity to give thanks for a safe journey on arrival, or to pray for one on leaving. After the Reformation, when it was damaged in battle, it suffered a mixed heritage; it has been a jail, a dwelling and a tobacconists, until 1913 when it was bought by Sir Charles Stoddart. It was eventually rededicated in 1924, when it was returned to its original use. A new bridge was built in 1930 when the old bridge was restricted to pedestrians. It is one of the very few surviving Bridge Chapels in the country. Further restoration was completed in the 1970s when a new window was installed.

Disasters are never far away from rivers, and one such was recorded by Anthony P. Munford in his *Rotherham – A Pictorial History*. In September 1931, the petroleum barge *Michael* was swept over the weir until it came to rest on the new bridge, necessitating the use of winches and steam rollers to haul it clear.

The town is well provided for with markets and shops, but its proximity to the Meadowhall shopping centre means it does not have the wide range of shops found in other large towns and cities. It was the victim of an economic crisis when some of its steel industry declined and the Burberry clothing factory closed in the 1980s, but has benefited from some regeneration funding in recent years. It's not without a sense of fun either, holding 'a day at the seaside' event in 2011.

Foundries along the river produced high quality goods from the early 1800s, including cannons for HMS *Victory* and cast-iron bridges. Steel, Peech & Tozer's massive Templeborough steelworks, reputed to be a mile long and employing 10,000 people at its height in the mid-twentieth century, became part of the British Steel Corporation and eventually closed in 1993. Magna, the Science and Adventure Centre, now tells its story.

Following the 2007 floods, which affected this area, a new wetland and flood storage area was created at Centenary Washlands by the Environment Agency and Rotherham council, which Sheffield Wildlife Trust manages as a local nature reserve.

Steel Peech & Tozer as it used to be, though looking much cleaner than it did in the middle of the twentieth century, courtesy of John Reynolds.

Magna as it is today.

3

NAVIGATION

As will have been noted from previous sections, navigation of the River Don was impossible due to the shallowness and the fall of the water as it tumbled down from the hills and moors of the Pennines through reservoirs and valleys, villages and towns. But, as early as 1698, an Act of Parliament was presented by Sir Godfrey Copley of Sprotborough, MP for Thirsk, representing the interests of Rotherham.

The River Don was recognised as being a cheaper and faster way of transporting imports and finished goods from and to sea ports, but there was opposition from land and mill owners who feared they would lose their investments. Town councils also had their own objections, and it wasn't until 1722 that a partial agreement was reached between these two sectors when Sheffield agreed that it would be responsible for the river to Doncaster if Doncaster would be responsible for it downstream from there.

However, it was not until 1726 that it was agreed that several cuts could be made, and that the river could be made deeper and wider, which would enable boats of 20 tons to reach Tinsley from the Humber. Restrictions still remained to protect some privately owned water-powered installations, such as those belonging to Sir Godfrey Copley at Sprotborough. In 1727, Doncaster Corporation sought powers to make further improvements and the Bill was passed uneventfully.

Needless to say, this was still only the start of the process, as it wasn't until 1819 that it was possible to sail directly from the Humber to Sheffield, and many years were to pass before negotiations and Acts of Parliament enabled construction of the canal system we see today.

Following many ups and downs, the last major attempt to revitalise it as a working system was made in 1983, when it was upgraded to a 700 ton Eurobarge standard by deepening the channels and enlarging the locks as far as Rotherham. Unfortunately, the wished for rise in freight traffic failed to materialise, and there is currently only one tanker plying its trade up and down the waterways, leaving them available for leisure activities.

The waterways changed hands several times over the years, but in 2013, the British Waterways Board – the government organisation that had eventually taken control became The Canal & River Trust, a charity that is responsible for its own fundraising.

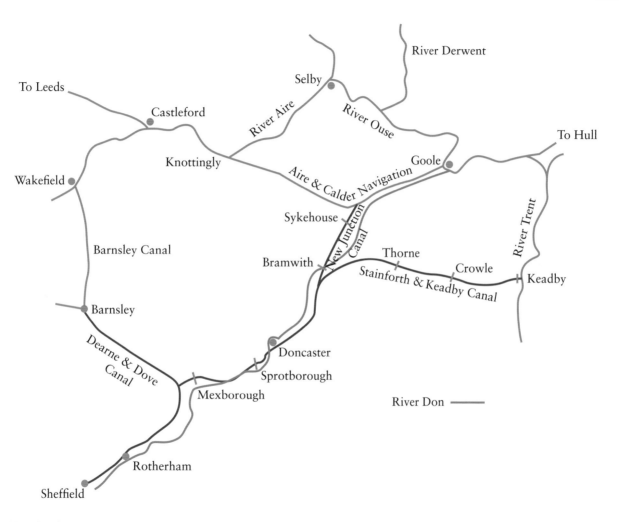

River Derwent

Selby

River Aire

River Ouse

To Leeds

Castleford

To Hull

Knottingly

Goole

Aire & Calder Navigation

Wakefield

River Trent

Sykehouse

New Junction Canal

Barnsley Canal

Thorne

Bramwith

Crowle

Keadby

Stainforth & Keadby Canal

Barnsley

Dearne & Dove Canal

Doncaster

Sprotborough

Mexborough

River Don

Rotherham

Sheffield

The Sheffield and South Yorkshire Navigation.

VICTORIA QUAYS TO THE HUMBER

Sheffield Canal Basin, now known as Victoria Quays, dates from 1814 and enabled canal boats to reach the heart of Sheffield from the Sheffield & Tinsley Canal. It served the city well for many years, with its enormous warehouses, but the railways gradually usurped its role and it declined into a rather sad and neglected state. Efforts were made to revive it in the 1990s, when hotels were built in the vicinity, the Sheaf Works were turned into a pub and the derelict railway arches became shop units.

There are good bus, Supertram and rail links nearby, but it has still not yet achieved its potential and I fear more money will be needed to make it the tourist destination that it could be.

As the canal here is not connected to the River Don, I was curious to know from where it obtained its water, and was informed that it is pumped from the river near Meadow Hall and the Tinsley Viaduct, where the canal and river run parallel to each other until they converge beyond Jordan's Lock. This was not the whole story, however, as it was also fed by mine water from a colliery owned by the Duke of Norfolk, which resulted in the canal being bright orange in colour due to the iron oxide contained in the discharge water. I am pleased to say this was not the case when I visited.

Victoria Quays has many interesting buildings and narrow boats moor here too.

Above: The Straddle Warehouse.

Below: The railway arches.

Above: Merchants Crescent and grain warehouse.

Below: The river and the canal run parallel to each other as the Tinsley Viaduct passes overhead.

The Sheaf Quay.

Beyond the Viaduct, the Don Navigation utilises the river where passable, but weirs and convenience make canals necessary for long stretches. There are several locks to negotiate before reaching Rotherham, and Swinton, Mexborough, Sprotborough and Doncaster locks are all encountered before arriving at Bramwith. Here the waterway divides into two: the left-hand stream becoming the New Junction Canal, which joins the Aire & Calder Canal at Southfield Reservoir; the right-hand stream being the Stainforth and Keadby Canal, which feeds into the River Trent.

ALDWARKE AND EASTWOOD

I had to look Aldwarke and Eastwood up on Google Earth to find out where they were as I thought they were in an area that was totally unknown to me. 'Go down the A630 for 7.2 miles from Warmsworth', it told me, 'and, at the roundabout, take the third exit'. When I arrived at the roundabout, I recognised it as having previously been known to me as the location of the famous 'mushroom garage' (easily recognised because of its design), which disappeared a good number of years ago now.

Turning right onto the B6123 as instructed, I almost immediately crossed two bridges and a small road on the left, giving access to Aldwarke Waste Water Treatment Works, which belongs to Yorkshire Water.

Above: The first lock after the Tinsley Viaduct.

Below: The road bridge viewed from above Aldwarke weir, and a favourite fishing spot below the weir.

on. It was not very clear, but it would appear that Wash Lane bridge gives access onto the 'island' between the two waterways.

Crossing the B6123 to the other side of the road, I was surprised to see Aldwarke Lock, right next door to the Asda supermarket.

Left: Aldwarke Weir.

Right: Wash Lane Bridge.

Parking on this road and climbing up a bank, I found myself on a towpath, with the road bridge and a weir to my left, and quite a sharp bend going away from me on my right. This was obviously the river.

Returning to the main road, I turned right to look over the first bridge, where I was met by the sight of the weir, and, continuing on to the second bridge, discovered a view of Wash Lane Bridge. This is the canal, and the sharpness of the turn once one has travelled beneath this bridge can be seen by the angle of the safety barrier above it, which bars access to the river and the weir immediately to the right, and the canal continuing straight

Aldwarke Lock.

Having found Aldwarke, my next destination was Eastwood Lock. But where was it? Several enquiries bore no fruit, so, taking a chance, I continued past the water treatment works to a cul-de-sac. Behind a big gate was Eastwood Lock with a wharf/marina alongside, which, I discovered, was part of the Waddington domain, of which more later. Fortunately, Martin let me in.

Being situated where it is, there is no doubt of Eastwood's place within the industrial heritage of Rotherham, but its location within the navigation of the River Don cannot be neglected either, as it played an important role in its later history too. Prior to 1966, when a scheme to improve a section of the Sheffield & South Yorkshire Navigation was submitted to Barbara Castle MP, the then Minister of Transport, there was a desire to develop this waterway, thereby increasing the volume of raw materials and goods being transported to and from Europe in the cheapest way possible. This scheme was rejected, and it was not until 1973 that a new proposal was considered by the government. This Bill received Royal Assent in 1974, but the money to realise it still had to be found. It wasn't until April 1979 that finance had been acquired from the government and an EEC grant, and the first pile was driven in at Pastures Road Bridge, Mexborough. At the same time, a plaque was unveiled by The Rt Hon Peter Shore, the Secretary of State for the Environment, to commemorate the beginning of the biggest redevelopment scheme on Britain's waterways for seventy years.

Eastwood, near Rotherham, was to be the terminus of this scheme, which would enable 700 ton barges to arrive there from the Humber following the straightening, widening and deepening of the channel of the River Don and the lengthening of ten of its locks.

It was the last lock to be completed, and it came into use at Christmas 1982, six months before the completed upgrading was officially opened by Sir Frank Price, the chairman of the British Waterways Board, on 1 June 1983, when Eastwood Lock was renamed the Frank Price Lock in his honour.

As I have noted elsewhere, it was all too late. Commercial traffic declined and the main beneficiaries today are those who use the navigation for leisure purposes. Sadly, the diamond-shaped plaque renaming the lock also appears to have disappeared, leaving only a vague shadow as a reminder of the dream that faded away. Nevertheless, the lock lives on.

In the middle distance of the photograph (on the top left of page 39) a footbridge that crosses the canal to the Parkgate area of Rotherham is just visible. The footbridge is considered unsafe at present, but a new one is expected to replace it in the near future, giving access to the Parkgate Shopping complex.

Leaving Eastwood, I returned to the B6123 at lunchtime to taste the delights of Greggs bakery at Parkgate before returning to the A630 and Hooton Roberts and Kilnhurst, where one crosses the river and the canal runs alongside the railway.

However, while en route, I saw a sign for a transport museum and decided to call in. It was closed, but I was invited in to look around. It is mainly devoted to buses, but had a small exhibition about the River Don and is worth a visit.

It's not far from here to Swinton.

The canal arrives from the Victoria Basin and Rotherham and departs for Aldwarke, along an attractive stretch of water.

The river and the canal at Kilnhurst.

Above: A Waddington's boat.

Below: Swinton Lock.

Right: Waddington's boatyard at Swinton Lock.

Above left: I was told the holes in the wall were drilled in readiness for the bridge to be blown up if ever the enemy landed during the Second World War.

Above right: The railways run next to the canal.

Opposite left: A right-hand bend can be difficult to navigate.

SWINTON, MEXBOROUGH AND DENABY MAIN

These three small villages, now almost running into each other alongside the Don, grew as more and more people were drawn to work in the industries that spawned there.

Swinton was once an internationally renowned centre for the manufacture of ceramics; Rockingham Pottery made world-class porcelain until 1842. Other industries included glassmaking, engineering and deep-coal mining.

Victor Waddington is recognised as the 'Giant of the South Yorkshire Waterways' in a book by Mike Taylor, being a designer and builder of canal barges, as well as a philanthropist. Beginning his business in Mexborough, he soon moved to Swinton and later opened a wharf at Rotherham. He was responsible for building barges and carrying much of the river trade in them, and various members of his family joined the business to work with or for him. He died in 1999, aged ninety-one, disheartened with the way the canals were going, but hoping his two sons might be able to carry on; the boatyard, with a few boats alongside, remains near Swinton Lock, which is known locally by his name. The small town of Swinton now suffers from the loss of its major work providers.

The lock at Swinton is the same size as that at Sprotborough, being similarly enlarged in the 1980s.

Leaving Swinton, two railway bridges, one derelict and one still in use, stand side by side as the river progresses towards Mexborough; drill holes can be seen in one of them, evidence, I was told, that it could have been blown up had there been an invasion in the Second World War.

Mexborough is just around the corner of the river from Swinton, and was built on similar industries. In its early history though, there was an Iron Age settlement here, and the Romans are known to have had a fort or villa sited on land between Doncaster Road and Pastures Road.

Station Road Bridge is next to appear, and a wide swing is necessary in order to manoeuvre a boat through it, turning at a right-angle as it does before a massive building appears on the north bank.

This was originally a flour mill owned by James White; the Don Roller Mills as they were then called, were later taken over by Barnsley Co-operative Society in 1912.

More recently it has been used for making products out of mild steel tube and wire for the education, retail equipment and industrial sectors.

A towpath is a reminder of the time when horses pulled the barges, and a tunnel can be seen alongside one bridge which provided a pathway for the horses when the towpath ran out. The bargees would undoubtedly pole the boat through the bridge, hooking up with the horse on the other side before carrying on with their journey.

The homes of the early inhabitants of Mexborough appear to have clustered around the ferry, where there was safe passage across the River Don, even when in spate. With protection from a high cliff, an abundance of clean drinking water for both livestock and residents and plenty of fish to eat, J. R. Ashby tells us that the town thrived in *A Short History of Mexborough*, in time building substantial stone houses and making pots. Ferry Boat Lane was obviously named after it, and the crossing lasted there until 1963.

As the best point for crossing the river, travellers would have

The Old Flour Mill and some early employees, courtesy of Mexborough Heritage Society.

Above: A tunnel to enable horses to bypass the bridge.

Below: Station Road Bridge.

Opposite right: Mexborough Top Lock.

A footbridge enables pedestrians to cross the canal and the river, with the 'rapids' that make the lock necessary.

found it a convenient route for transporting goods, hence the attraction for settlers.

Mexborough Top Lock can now be found at the end of Ferry Boat Lane, off Church Street, bypassing the shallow 'rapids' on the river that runs alongside. A footbridge now crosses both the canal and the river here.

Also on Church Street are three almshouses. Originally, six were gifted for poor widows by William Horne of Mexborough Hall in 1669 and built on Market Street, but they were relocated due to road widening in 1909. However, expectations regarding the size of homes have changed over time and they have now been converted into three larger dwellings.

Mexborough today is quite a busy small town, boasting a hospital, a library and shopping precinct where most needs can be met.

It is proud of the fact that Ted Hughes, the poet, and Brian Blessed, the actor, grew up here, and it is also the birthplace of 'Iron' Hague, whose headquarters were the Montagu Hotel, and who became the heavyweight champion of Britain in 1909 when he beat 'Gunner' Moir in 2 minutes 47 seconds – a disappointment for all those who had hoped for a longer spectacle.

The cemetery contains a memorial to 'Iron' Hague, and there is also a gravestone in St John's churchyard for three victims found in Mexborough following the catastrophic flood that occurred on 12 March 1864. When the dam of the Dale Dyke reservoir near Low Bradfield, above Sheffield, collapsed, its waters flowed down through Sheffield and beyond, killing 250 people.

Denaby Main's roots lay in the mining industry, but the grime of its past has long gone. New homes have now been built and the Dearne Leisure Centre sits on the site of the colliery.

Cadeby Colliery, which was sunk alongside Denaby Main, has also gone, but the legacy of these pits remains in a memorial to the 203 men and boys who died in them between 1866 and 1968. Almost half of these deaths were caused by explosions that shook Cadeby Colliery on 9 July 1912. During the rescue operation following the first of these, a second explosion occurred, leaving a total of ninety-one dead.

On the evening of that day, King George V and Queen Mary, who were staying at Wentworth Castle at the time, visited the colliery to pay their respects; it was reported that had some workers not taken an unofficial day's holiday for the royal visit, many more might have died.

As the river flows on its way to Conisbrough, it passes the site of Cadeby Colliery, which became the home of the doomed Earth Centre. This was a project to encourage sustainable development that was opened with great enthusiasm in 1999, but closed after going bankrupt in 2004. It is now an educational adventure centre.

Left: Almshouses plaque.

Right: Mexborough's shopping precinct.

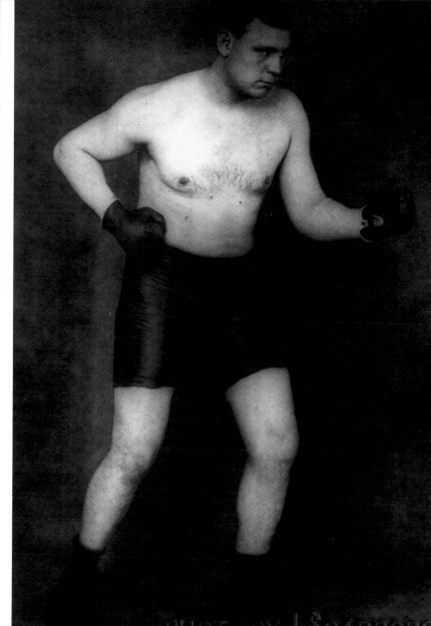

Above: The Montagu Arms, location of 'Iron' Hague's training sessions.

Right: 'Iron' Hague, courtesy of Giles Brearley.

Above: The Earth Centre was built on the site of Cadeby Colliery, but is now an education facility.

Right: The Dearne Leisure Centre.

Opposite left: Denaby Main Colliery, courtesy of Mexborough Heritage Centre.

CONISBROUGH

The Norman keep of Conisbrough Castle stands impressively on the hill top just to the south of the river. It would have been easily visible when Sir Walter Scott stayed at what is now The Boat Inn at Lower Sprotborough, around 3 miles downstream. It is reputed that he wrote a chapter of his famous novel while visiting the area and, as a result, the castle is known throughout the world as the 'Saxon' fortress in *Ivanhoe*.

The castle was built around 1070 by William, the first Earl Warenne, son-in-law of William the Conqueror. Although he was given property in many different parts of England, his principal holding appears to have been his Yorkshire Estate, of which Conisbrough was considered to be the head. From this, it can be seen that Conisbrough was a very important stronghold, and in fact the name is derived from the Anglo-Saxon word Cyningesburgh, meaning 'the defenced burgh of the King'. At the time of the Norman Conquest in 1066, it was held by King Harold, who was defeated at the Battle of Hastings.

The view from the top has changed considerably over the years. What was once a deer forest and hunting ground was replaced by railways and coal mines, while arable fields and woodlands became housing estates as the Industrial Revolution progressed. The twenty-one-arched viaduct that straddles the river is visible in the distance, as is St Peter's church, believed to be the oldest building in South Yorkshire, dating from the eighth century and forming the centre of this large and important parish.

Above: Conisbrough Castle.

Below: Conisbrough Viaduct.

The river obviously played an important part in this history and, in its heyday, was a busy link in the chain, with all manner of products from coal, timber and limestone to petrol and wheat passing through its lock in a variety of vessels: keels with sails and Tom Pudding tugs or barges, which sometimes towed compartment boats. Barge captains managed the boats, often employed by fleet owners, but whole families lived on them too. During the 1950s, the only horses left pulling the barges were on the internal coal trade. Before the arrival of motorised barges they were a regular sight, though it was not unknown for people to bow-haul or pole their boats either through difficult obstacles or when the sails lacked wind. If the towpath changed sides, it would then be necessary for the horses to be carried across the river on flat-decked chain ferries. Stables were located at appropriate points along the route where the horses could rest, looked after by horse-marines, who were known for their black trilbies, mufflers, corduroy waistcoats and narrow trousers, with massive leather belts to hold the horse whips completing the outfit.

Another important industry associated with the river is the cannonball factory, which used the mills at Burcroft. This was owned by Walkers in the seventeenth century, and a painting by Turner depicting this is now held by the Tate Gallery. However, in 1855, Geo Booth & Son established its business there, and by the twentieth century it was producing sickles, reaping hooks and grass hooks that were exported to almost every country in the world. This factory, too, has now disappeared, but a photograph shows two young women and a supply of sickles laid out on tables. The lady in the centre is Audrey Emerson, who lived at the lock house with her parents and is shown in

Above: The view from Conisbrough Castle.

Below: Gordon and Audrey Mead.

Right: Audrey is ferried across the river to church on her wedding day from Conisbrough Lock House (*above*), courtesy of Gordon and Audrey Mead.

Opposite left: River view.

another photograph crossing the river in a boat on her way to church to be given away by her uncle on her wedding day.

Gordon and Audrey, now in their eighties, who began their married life living at the lock house with Audrey's mother, remember working there well and have a fund of stories to tell of life on and alongside the canals.

In the 1980s, several locks along the river were enlarged in the expectation that Rotherham would become an inland port, its commerce stretching across the English Channel to the continent in linked container barges, but by the time the scheme was complete, the steel and mining industries had been reduced to a shadow of their former importance, and the scheme was curtailed. However, as part of this plan, the lock at Conisbrough was removed and the height of the weir at Sprotborough was raised.

Visitors to the town today might be drawn to the castle, restored by English Heritage in 1992, and by the new visitor centre, but there is a steep hill to negotiate if they wish to see the river, which might deter the less able.

'I wouldn't touch it with a barge pole!'

This saying relates to the pole used by bargees to push the barge away from the river bank or the side of a lock if they got too close. Poles, about 10 feet long, had a spike and a hook at one end, thereby helping to push the barge away or propel it along.

It usually means that one wants to keep away from something or someone as there is something dishonest or unreliable about it/them that might tarnish one's reputation.

Above: The sickle factory, courtesy of Audrey Mead née Emerson (seen in the centre).

Opposite right: The Don Gorge.

Conisbrough Castle projection, courtesy of English Heritage.

LEVITT HAGG AND LOWER SPROTBOROUGH

The entrance to this area of outstanding natural beauty is marked by the twenty-one arches of Conisbrough Viaduct. With fourteen arches on the north side and seven on the south, it formed a connection between the Lancashire & Yorkshire Railway and the Great Northern and Great Eastern Railways, carrying passenger trains over the Don until 1951. Having survived the possibility of demolition in the 1980s, due to the presence of dwarf elder (*Sambucus ebulus*), the viaduct now forms part of the Trans Pennine Trail. Dwarf elder or not, it would have been a travesty to have destroyed this important part of the gorge's heritage.

It was a great feat of engineering in 1906/07 when it was built by Henry Lovatt Ltd, being 1,584 feet in length and containing 15 million bricks. The section over the river was accomplished by the use of a cradle, known as a 'blondin', and it is said that each arch was built by a different subcontractor, no single company being big enough to complete the whole of the viaduct.

Although the viaduct is no longer used by trains, the main railway line from Doncaster still runs along the south bank of the river, until it crosses the Don, before heading towards Denaby and Sheffield.

Dwarf elder is not the only interesting plant species found in the gorge, with flamingo moss being of particular note and a patch of cotton grass, seemingly quite out of place in this magnesian limestone area.

There is no doubt that the Don Gorge, as it is known, could merit a book in its own right, with its abundance of historic features. As might be expected in a limestone gorge, quarrying has played an important part in its past, though those quarries situated by the river side have now closed. Before it finally reached this demise in 2013, however, Tom Puddings were regularly used to ferry limestone down the river from Cadeby Quarry to Hexthorpe, to create the base for a new housing estate being built on the old railway plant works and athletic ground. The only company remaining is one that produces polished stone for exclusive building projects, and the *Humber Princess* is the only super-tanker still plying its trade up and down the Don, though the smoky tug shown overleaf was carrying stone to Hexthorpe a few years ago.

Continuing along the river, it is not long before one reaches the 'lost' village of Levitt Hagg, unfortunately now a landfill site. Located on the south bank, it was a thriving community from around 1750, a row of white houses being built in 1815 and a mission hall in 1878. Boatbuilding occupied residents from around 1868, but the last boat was launched in 1901 when children were given a day off school to watch the end of this era.

The village, however, was mainly based around the large quarry and lime kilns, known as the 'Levitt Hagg Lighthouse' due to the glow which could be seen for miles around when lime was being burnt. These existed until the mid-1950s, when the quarry was closed and most of the residents moved into new council houses, which were being built at Warmsworth and Sprotborough after the Second World War. True, the houses were substandard and sometimes flooded, but it is regrettable that some part of this village's heritage could not have been saved rather than being used as a landfill site.

The hope that a project to inform visitors about the lime kilns, parts of which can still be seen, remains, but, as always, lack of finance is the inhibiting factor.

Above: A smoky tug.

Below: The Humber Princess.

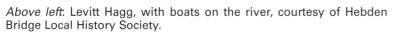

E.L.S. 60-5. Levitt Hagg.

Above left: Levitt Hagg, with boats on the river, courtesy of Hebden Bridge Local History Society.

Above right: Mission Room, centre right.

Below: Remains of part of the lime quarries today.

Above: Levitt Hagg.

Left: The Boat Inn. Note how the pine tree now towers over it.

The picture (*opposite*) shows Levitt Hagg, with the quarry face behind it, and Lower Sprotborough in the foreground. The fields shown to the right, now the site of the Flash nature reserve, suggest it was taken prior to 1924 when the mining of coal at Edlington caused subsidence and the formation of a lake, three quarters of a mile long, running parallel to the river. The nature reserve was opened in 1984, as a joint venture between Doncaster Council and Yorkshire Wildlife Trust.

At Lower Sprotborough, the Boat Inn now benefits from the attractions of the area, which draws visitors from far and wide.

From pointers gained during the renovation of two of the six cottages (*shown opposite*) located beside the pub, it is believed that Nos 3 and 4 date from the seventeenth century, having been converted from a barn. Nos 2 and 5 and the semis at the top, known as Tower Cottages, are thought to have been built in the 1850s. Water for these cottages was obtained from a well at the top of the lane until the 1930s, after which a bowser was filled on a weekly basis until a water supply was connected. The well has now been sealed off and a cairn constructed by the Don Gorge Conservation Volunteers contains a time capsule that tells this story.

A walk past these eye-catching cottages in a westerly direction brings one to a view of Sprotborough Flash, the stretch of water created from subsidence due to mining in the early part of the twentieth century, and is now managed by Yorkshire Wildlife Trust. By continuing along this path, it is possible to do a round tour, arriving back along the river bank at The Boat Inn. A much longer walk can be undertaken following this path, through the ancient Pot Ridings Wood, over Conisbrough Viaduct and

Above and below: Lower Sprotborough then and now.

Right: Snowdrops lead the eye to Sprotborough Flash Nature Reserve.

Opposite left: Volunteers take a well-earned break.

Taken from The Flash,
St Mary's church rises
above the trees.

Above: The canal and the river bridges.

Below: Another view of the bridge from the riverside.

Above: The toll house.

Below: The rectory.

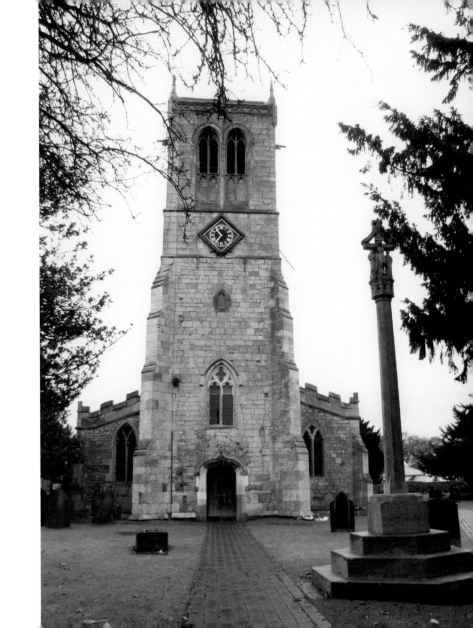

St Mary's church stands in the centre of the village.

The old fulling or flint mill, courtesy of Hebden Bridge Local History Society.

returning through Levitt Hagg. The Flash attracts many species of waterfowl and birds, and a meadow area is renowned for its cowslips and orchids that especially like the poor limestone soil.

The Copley Estate encompassed the whole of this area until it was sold in 1926, following the deaths of the lord and lady of the manor within the same week, drawing double death duties. The Boat Inn was one of twelve farms within the estate, and has had several incarnations since it was built in the 1600s, as Ivanhoe House, the Ferryboat Inn, the Copley Arms (which can still be seen carved in stone on its frontage) and Boat Farm. As mentioned in the previous chapter, it now delights in the possibility that Sir Walter Scott wrote a chapter of *Ivanhoe* while staying there. The age of the Scots Pine tree in the front garden is unknown, but it is probably approaching 200 years and, as can be seen, is now over double the height of the Boat Inn itself.

The river divides at this point, following the creation of a canal and cut that enabled a course past the weir. However, a new and bigger lock was built in 1831, and again in 1980, following much straightening and piling of the canal to enable even bigger barges to navigate their way to Goole.

The removal of the lock at Conisbrough and the raising of the height of the weir at Sprotborough has resulted in the canal being much deeper now than in the past, putting The Boat Inn and the cottages at a far greater risk of flooding. This danger has been partly alleviated by the installation of pumps and the design of the island between the canal and the river, which enables the water to overspill from the canal into the river below the weir when in spate. However, it didn't save them in June 2007 when they were inundated by over a metre of water.

Bridges at Doncaster and Sprotborough are the only available river crossing points in the area. Consequently, the two at Sprotborough, which are quite narrow, are extremely well used for access to Doncaster and the A1(M) motorway, which crosses the river further downstream. The river bridge was erected by Newton Chambers & Co. Ltd of Thorncliffe Ironworks in Sheffield, in 1897. It is due to undergo renovation in September 2014 when one-way traffic will be necessary.

On the corner of the canal bridge stands the square, flat-topped toll house, lived in by a local artist. As its name suggests, a charge was originally made to cross the bridges that replaced the ferry, which was previously located near The Boat Inn.
Should one wish to deviate a little from the river, one should go up the hill from here to the old village of Sprotborough, where the Norman church of St Mary takes pride of place. En route, you will pass the Old Rectory where Group Capt Sir Douglas Bader reputedly misspent his youth.

In the opposite direction, going over the two bridges, you will find yourself going up the hairpin bend. There is an access road to Levitt Hagg, with the fish and eel pass on the right (though it is blocked to motor traffic), and at the top, Warmsworth village is in sight. Old Warmsworth is entered by crossing the A630 at the traffic lights, and immediately turning left in front of the community library.

There are known to have been two water-powered mills in the Don Gorge. The foundations of the mill on the south side of Sprotborough Falls were exposed during the installation of the fish and eel pass in 2013/14, another link in the chain that enables fish to reach their spawning grounds upstream. Such a good job has been done in cleaning up the river in the past fifty

Dennis Petty proudly holding his trophy, courtesy of Dennis Petty.

Above: Sprotborough Lock where the canal joins the river once more.

Right: Open Day at Sprotborough Lock.

The ruins of the pump house.

The balustrade can just be seen on the ridge.

Sprotborough Hall, courtesy of Hebden Bridge Local History Society.

years that salmon have been seen, though it may be some time before they can be caught on anything like a regular basis.

This mill was initially a fulling mill, where woollen cloth was felted, either by feet or machinery, before being dried on tenterhooks and made into garments. The mill was also used to grind flint for the pottery industry at Swinton, and wheat for flour. The land adjacent to it was known as the camping ground and photographs show visitors picnicking and enjoying themselves there. It is hoped that the presence of the fish pass will result in this long-neglected area being brought into use once more.

A charter dating back to 1279 recorded a corn mill on the island between the river and the canal, opposite the present landing stage where the *Wyre Lady*, a pleasure boat, picks up its passengers. Remains of the building could still be seen until the 1960s, but a great deal of landscaping has taken place in the intervening years.

The flour mill stood on the island, the miller living in one of the cottages across the canal, but these were demolished in the mid-twentieth century. It is believed that the arches on the left could be a small lime kiln as there was evidence of a small quarry in that location until it was filled in.

The lock was manned for many years, the keeper living initially in a lock house alongside, but, due to the many changes made to the canal and automation of the lock, the practice of employing a live-in keeper was discontinued.

It was the first lock to be enlarged in the scheme to open the waterways to 700 ton vessels. The lock-keeper employed at that time was Dennis Petty, who went on to win the Castleford Area Best Kept Lock Award for four years running, as well as becoming National Champion in 1984. New gates were installed in 1996 at a cost of £26,000 when Dennis's son, Nigel, was lock-keeper. In February 2014, gates at the eastern end of the lock were replaced as part of a £100,000 maintenance operation and, on this occasion, I was one of the 1,800 visitors who took the opportunity to climb down into the 7.5 metre deep lock and walk its length. Here I was surprised to find freshwater mussels attached to its walls.

On the left of the photograph on page 77, stone arches, which enable water to escape onto the fields in times of flood, can just be seen. Moorings are now provided on the canal side of the island for overnight stays by boat owners.

Just past the lock, the remains of Sir Godfrey Copley's pump house and water engine can be seen. This enabled water to be pumped up to the village, but also to the roof of Sprotborough Hall to supply a fountain that rises to 290 feet, second only to the Emperor Fountain at Chatsworth House. Sprotborough Hall, described as a Charles II mansion, was built in the 1670s for the Copley family, 100 feet up on the escarpment above the river. The pump house also supplied a lead-lined swimming pool, 10 metres by 5 metres by 1.8 metres deep, in the grounds. The hall was unfortunately pulled down after the sale of the estate in 1925, and the only thing remaining is a balustrade, which can be seen most clearly in winter when there are no leaves on the trees, and the ruins of the pump house.

The gorge is brought to an end by another more modern structure, the A1 motorway bridge, after which the river meanders eastwards on the next stage of its journey.

From the above descriptions, it will be recognised that despite the loss of physical evidence, this stretch of the River Don has a wonderful heritage, and, along with several areas of outstanding natural beauty that have SSSI status, is one that is valued and loved by many.

Members of the Don Gorge Community Group and its conservation volunteers have given many hours over the past two decades to preserving and conserving the natural beauty of the area, as well as to educate locals and visitors in responsible tourism; credit must be given to them and those agencies that support them in their endeavours.

When John Holland, who you will remember found the source of the Don in 1836, arrived in Sprotborough on his tour of the Don, he was moved to write a poem called *Morning Walk to Sprotbro*, recording his impressions of this beautiful area:

> O, happy Sprotbro! spot
> Romantic, where sweet nature's charms salute
> The roving eye! The church, the mansion, and
> The lowly cot, delightful vales by hills
> Encircled round – I fain would ask, who has
> Not been to Sprotbro'? Let them go – 'twill well
> Repay them for their toil, if such a toil
> It be, where pleasure leads the way: If they
> Have eyes to see the charms and beauties of
> The scene, as mine beheld them once when I
> Strayed thither: and if they have hearts to feel
> The pure delight that such sweet sense inspire –
> Then let them go.

It is still well-regarded today, attracting many visitors and being recognised as the 'jewel in Doncaster's crown'.

Two more railway bridges remain before Doncaster is reached, one still operating and the other now used by walkers and cyclists, all three of which can be seen in the photograph below.

The hamlet of Newton with its two farms is the next focus of habitation on the north side of the river, but this section of the river is actually an artificial channel dug to improve navigation. Originally, the river meandered across the meadows, a feature that I am informed can be clearly seen from aerial photographs.

Hexthorpe, which is near here but on the south bank, was renowned for its park and 'Dell' in the 1950s and '60s, when thousands of people would flock to see its flower displays and aviary to the music of the Dagenham Girl Pipers.

Golf and putting were also available. I believe the Dell was created out of an old quarry where stone tiles called 'flatts' were made to roof houses, hence the park being known as Hexthorpe Flatts. Today, the Dell is a shadow of its former self, though many people will have fond memories of times past. Doncaster Rowing Club, which still operates on the river, has its base here.

Above: Travelling west from Doncaster.

Below: The Dell.

Sunset over Lower Sprotborough.

Doncaster Lock and passengers wait for the *Wyre Lady* for a river trip.

DONCASTER

Located on the main A1 road between London and Edinburgh, Doncaster, or Danum as it was known to the Romans, was for many years renowned for its butterscotch and its mintoes. The word 'butterscotch' was first recorded in Doncaster, where Samuel Parkinson began making it in 1817. Buying a tin of Parkinson's Butterscotch, which had the royal seal of approval, was one of the highlights of the many visitors to the St Ledger race meeting in September. Nuttall's Mintoes were equally famous and, though I understand that they can still be bought, I wouldn't like to guarantee that they taste the same.

Today, St George's church, now a minster, and The Hub, the new college building, sit each side of St George's Bridge. Built in 2002, it spans the River Don, the South Yorkshire Navigation Canal and the east coast mainline all in one go.

The new college waterfront provides a very pleasant seating area overlooking the canal. The marina, which provides moorings for narrow boats, is only a short walk from the famous Doncaster market and shopping centre. Around 1 mile further downstream, Strawberry Island has also been developed as a boat club.

The River Don arrives in Doncaster near the prison, which replaced a power station in June 1994, and is bounded by two rivers the Don and the Cheswold. The latter, possibly the shortest river in England, feeds into the Don after intersecting Marshgate, which is situated under the Great North Road bridge on its way north.

This area suffered from regular flooding until 1846. Arches had been built further along the Great North Road to allow

Toy model of a Nuttall's Mintoes iconic delivery van, courtesy of Oxford Diecast.

MADE IN DONCASTER

NUTTALL'S MINTOES

A DELICIOUS COMBINATION OF SUGAR, TREACLE, BUTTER AND OTHER EDIBLE FATS

8d QUARTER POUND

PARKINSON'S

SOIT · QUI · MALY

HONI · PENSE

DIEU · ET · MON · DROIT

CELEBRATED ROYAL

DONCASTER

BUTTER-SCOTCH

As supplied by permission (and to no other party was the same granted) to the

QUEEN & ROYAL FAMILY

ON THEIR VISIT TO DONCASTER IN 1851.

Also to Her Royal Highness

THE DUCHESS OF KENT

ON HER VISIT IN 1852

And extensively patronized by the NOBILITY, CLERGY AND GENTRY

PARKINSON & SON,

FAMILY GROCER

TEA DEALERS, CONFECTIONERS &c.
50 & 51 High St DONCASTER.
ESTABLISHED IN 1817.

The only genuine packets of Butter Scotch are signed

Above: Doncaster minster.

Below: The hub.

Opposite far left: Nuttall's Mintoes tin, courtesy of Market Lavington Museum, Wiltshire.

The flood relief channel, also known as the River Don.

River Cheswold.

Doncaster Mansion House.

water to escape, but on 6 August 1846, a cut was made in the Newton Bank between Sprotbrough and Doncaster, by men with no right to do so, in an effort to divert the water away from Marshgate. It was a success, but the men were brought before the bench for having taken matters into their own hands. Fortunately, the case was dismissed, the Corporation being blamed for not having cut the bank itself, with the added remark that the bank had probably been erected illegally in the first place. It is worth noting that when the Newton Bank was breached again in 2007, the flood water took the same natural course that it had taken in the nineteenth century.

The canal lock can also be found at Marshgate. It is not accessible under normal circumstances, but the photograph below shows members of the public waiting for the Wyre Lady to arrive to take them to the Don Gorge, for a special Park & Float event in May 2009.

The water below Marshgate Lock was originally the river, evinced by the number of oxbow channels between Doncaster and Long Sandall, with Strawberry Island being the most notable. However, the river and the canal split at around 800 metres above Marshgate Lock. What is left of the river course branches northwards to a sluice, which regulates the levels for navigation and flood management.

In 2000, when this sluice needed repair, the Environment Agency took the opportunity to create a fish pass in a new channel, bypassing the sluice, which enabled fish to migrate between the tidal and non-tidal River Don for the first time in more than 500 years. The pass was built to an innovative design in the form of a rock ramp, and was the first of its kind in this country, winning an Institute of Civil Engineers award for concept and design.

Below the sluice and fish pass, what is now regarded as the River Don flows under the North Bridge and is actually a flood relief channel, which was built in the early twentieth century to relieve flooding in the Wheatley area.

The flood relief channel eventually reconnects with the Don's natural channel close to Long Sandall, though it is very difficult to see where this occurs (I am indebted to Chris Firth for this information).

Despite the loss of its traditional confectionary, Doncaster is not without other claims to fame. The Mansion House, dating from 1751, was built as a place for corporate entertaining and has a wonderful staircase and ballroom. It is one of only three mansion houses built for this purpose, the others being in London and York.

The St Leger Stakes, which ran for the first time in 1776, continues to be run, with a new grandstand and improved concourse. The racing tipster Prince Monolulu was perhaps

Below: Sir Nigel Gresley Square.

the most colourful attraction until his death in 1965, his best-known phrase, 'I gotta horse!' being the name of his memoirs.

The railways also brought fame to Doncaster with the building of No. 4468 Mallard, which was designed by Sir Nigel Gresley and holds the world speed record for steam locomotives. It can now be found in the National Railway Museum at York.

Doncaster was also host to the renaming ceremony of the Blue Peter locomotive by the BBC Blue Peter presenters at a railway works open day in 1971, following its restoration at York, Leeds and Doncaster. Repainted in LNER apple-green livery as No. 532, its renaming was witnessed by 60,000 people. Long service followed, but it now awaits further restoration before its journey can be resumed.

Doncaster was badly hit by the miner's strike in the 1980s, and struggled to recover, but has remembered its railway hero by naming the site of the new civic centre and theatre in Waterdale in his honour.

STRAWBERRY ISLAND

Strawberry Island can be found at the bottom of Milethorn Lane, off Wheatley Hall Road, so-called as there was a thorn tree nearby, around a mile downstream from Doncaster town centre. The marina has been formed by an oxbow in the river as it left Doncaster, but this was bypassed by the canal to aid navigation. The island itself is used for allotments, but the water is used as mooring for pleasure craft. More about this can be found in the 'Leisure' section.

View from the club house.

LONG SANDALL

A little further along, Long Sandall comes into view, where there is a lock that can be used for the Sheffield class of boat, as well as the 700 tonner, by having an extra set of gates within it. Quite a lot of boats were moored here as well, the owner of one acting as lock-keeper. He informed me that he is a volunteer, as paid keepers are financially no longer a possibility, and that he had let twenty-two craft through that morning. Unfortunately, he wasn't even paid the penny for his labours that seems to have been the going rate for helping boaters through the lock in times past.

From here, I feel it is necessary to veer away from the exact course of my narrative to explain what happened in the seventeenth century. At that time, the River Don, as well as the rivers Torne and Idle, flowed through a very large tract of land known as Hatfield Chase and the Isle of Axholme, which flooded on a regular basis and was therefore very swampy. When the land was drained, the course of the River Don was changed dramatically and so I have inserted a section here to explain the course of the old river, before continuing with the new.

Left: The old River Don.

Opposite: Long Sandall.

LONG SANDALL LOCK

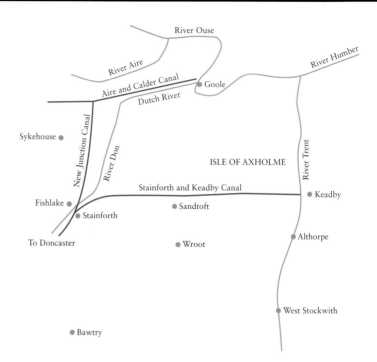

Above: The new River Don.

Left: This portrait was identified as being that of Sir Cornelius Vermuyden. However, on checking with Valence House, I was informed that Vermuyden was not recorded as being a member of San Marco and Lazaro, the chain and order of which the sitter wears. Furthermore, the sitter was once identified as Sir Richard Glanville, King Charles I's General of the West, and the portrait actually descended through his family for some time. Another complication arises as an identical portrait (painted by an unknown artist *c.* 1612) is currently held by the Usher Gallery, Lincolnshire. This one is identified as a portrait of Sir Philibert Vernatti (1590–1646), who was also involved in Fen drainage in the seventeenth century. On the balance of probability, it is more likely to be a portrait of him, with the confusion over the sitter's name resulting from the fact that they were both involved with Fen drainage. Courtesy of LBBD, Valence House Museum.

HATFIELD CHASE AND THE ISLE OF AXHOLME

In the seventeenth century, the Humber estuary was the collecting point of several rivers rising in Yorkshire and Lincolnshire, the River Don being just one. The river Aire came from the west to join the Ouse at Goole, and the River Don ran in a north-easterly direction from Stainforth, through Thorne and the Isle of Axholme to the River Trent at Ardingfleet; on the way it was also joined by the Idle and the Torne. The Trent, coming from the south, joined the Ouse near Faxfleet, from where it later converged with the Humber on its journey to the sea. The Isle of Axholme was so called as it was surrounded by these rivers and Hatfield Chase was the recipient of the overflow in times of flood.

The result was that the Chase was a very marshy area, though rich in wildfowl and fish and, as King Charles I was lord of the manors of Hatfield, Epworth, Crowle and Misterton, it was favoured by him as a royal hunting ground. Despite this, however, it was inaccessible at most times and, as he was rather short of money, Charles decided that draining the land would enable more profitable types of agriculture to be undertaken, which would swell his coffers.

A Dutch engineer, Cornelius Vermuyden (1590–1677), was gaining a name for himself in England due to projects he had undertaken on the Thames and Canvey Island and, having come to the notice of the king, was contracted to drain Hatfield Chase and reclaim land that was heretofore unusable. His reward for doing this work would be a third of the land created, another third reverting to the king, and the rest being for use by the local inhabitants. As might be expected, the latter did not care much for this scheme as they argued it relieved them of common land, which provided pasturage and peat for fuel; the people of Epworth had already lost part of their common land in the fourteenth century and did not want to lose any more. It was also generally believed that they would be left with the most unproductive areas. Law suits ensued, but little was achieved and the work commenced in 1626.

Many of Vermuyden's workforce were Huguenots, Protestants who were fleeing religious persecution on the continent. A settlement was established for them at Sandtoft. However, the resentment of the locals could not be contained, and both this settlement and the work being undertaken suffered attacks.

The scheme involved the diversion of the River Don northwards along the Turnbrigg Dyke to join the River Aire and the damming of the River Idle at Idle Stop, before rerouting it to join the River Trent at West Stockwith. A drain would be constructed to take the River Torne across the Isle of Axholme, from near Wroot to Althorpe on the Trent, and another would run from Idle Stop to Dirtness, near Sandtoft, where a third drain coming from the west would converge before entering another sluice at Althorpe.

Unsurprisingly, Vermuyden's work was not entirely successful and, because of the flooding of some villages that had previously been free of this problem, the ancient banks of Fishlake and Sykehouse were repaired in 1629 at a cost of £200, with any future repairs to be charged to the residents. However, by 1633, a new outlet was considered necessary and the Dutch River was dug in an effort to avoid future inundations. The cost of this new river was reputed to be £33,000 and the cost of the whole scheme was £400,000.

Unfortunately for Vermuyden, because of his errors, he lost most of the land he was due to acquire in the Chase. However,

Barnby Dun's lift bridge.

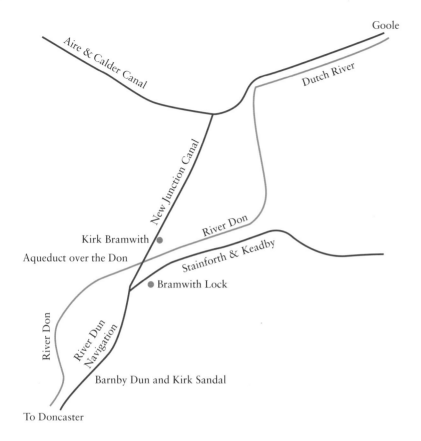

Above left: The routes of the New Junction and Stainforth & Keadby canals.

Above right: The beginning of the New Junction Canal on the left, with the Stainforth & Keadby Canal on the right looking towards Kirk Bramwith Lock.

Below: Bramwith Lock.

he was obviously able to overcome this loss, as he went on to buy 4,000 acres of land in Sedgemoor on the Somerset Levels, and Malvern Chase in Worcestershire. He also acquired an interest in the lead mines in Wirksworth, Derbyshire, which he drained by means of a sough. Vermuyden was knighted in 1629, and became a British citizen in 1633.

MS IN A RED BOX

When an unsolicited parcel was delivered to a London publisher, he was surprised to find that it contained a red box with the manuscript of a romantic novel based on Vermuyden's scheme within it.

No other information was included in the parcel, and so an advertisement was placed in two newspapers to try to ascertain the name of the author. This elicited a great deal of correspondence from readers, but no one claimed ownership. As seven of these correspondents suggested that the book should be published under the title *Ms in a Red Box*, the publisher decided to go ahead while there was still so much interest.

The novel is a romance, but also provides a great deal of background information regarding the way of life of the period and the area.

It is now available as a free download from the internet, or a print version can be purchased from the Isle of Axholme Family History Society.

Having explained the old river, we return to the current course of the river, which now runs from Long Sandall to Goole.

Above: The swing bridge on Low Lane.

BARNBY DUN

After leaving Long Sandall, the river and canal make their way towards Barnby Dun, whose name appears to describe a 'barn by the Dun'. The road across the canal at Barnby Dun is also the location of a lift bridge. I have only ever seen it raised once, and unfortunately the post van was in front of me, which rather spoilt the view. However, it was quite a sight and I include the photograph as a rarity, though the bridge must be raised regularly to allow the Humber Princess and other craft through.

Above left: Kirk Bramwith church.

Above right: View of the aqueduct from the river.

Below: View of the aqueduct from the canal.

A narrowboat passes under Stainforth Bridge.

The river runs parallel to the canal under another bridge.

KIRK BRAMWITH

Bramwith was something of a revelation for me. I have visited this area many times over the years, but never without getting lost. I lived at Fishlake on and off for a couple of years, and did eventually know how to get there, but this whole area is a criss-cross of narrow roads, canals and rivers, with very few road signs, making any journey a nightmare for the unknowing.

Having been guided to the river and canal at Kirk Bramwith by a friend and then left to my own devices, the first thing I needed to do was turn round and park, which was easier said than done. However, having found somewhere to turn round, I almost immediately missed a turning and found myself at Barnby Dun. Asking advice and retracing my steps, I suddenly saw a sign for Lock House Farm and decided to chance my luck down this private road; it was the best thing I could have done.

The farmhouse sits on the side of Bramwith Lock, which is on the Stainforth & Keadby Canal. Having found my way on to the towpath, I was able to take photographs and walk along it in a southerly direction for a couple of hundred yards when I found the second thing I was looking for, the place where the Dun (that's right, not the Don) Navigation splits into the New Junction Canal and the Stainforth & Keadby Canal (the Dun Navigation is a canal; from Doncaster the river bulges out to the west and then turns eastwards again to pass under the aqueduct before turning north to join the Dutch River).

My search was beginning to seem like a treasure hunt, as now I wanted to see the aqueduct on the New Junction Canal, but there was no sign of it from here. I now found myself in the midst of a heavy rain shower, so, rushing back to the car, I decided I would have to go back to the road, take two left turns on to Low Lane and cross the canal and the river again. Another piece of good luck came my way when I found the swing bridge over the canal being opened for a boat to pass through.

Eventually passing over it, I found a track going alongside the canal, but the aqueduct still couldn't be seen properly from here either. Now crossing the river, another private track appeared to go along the riverside. So, nothing ventured, nothing gained, I decided to try it. It led to a building belonging to the Dun Navigation, but a man appeared who had no objection to my climbing the bank to see the aqueduct in the distance.

As the aqueduct crosses the River Don, I also wanted to find the canal to see it from that angle. Fortunately, I was going in the right direction; a little further down the path, there it was. I was somewhat confused by the construction of the aqueduct, until I discovered that the large 'guillotine' gates, which tower into the air above it, are designed to be lowered when the Don is in spate to prevent the aqueduct being flooded, as well as the surrounding countryside.

This is the first of two aqueducts on this canal, as there is another one that that crosses the River Went near the end of its journey, at the Southfield Reservoir and the Aire & Calder Canal.

Parking is very difficult in this area as the roads are narrow and bendy, but I was able to stop for a few minutes to take photographs. Having achieved my aims for the day, I managed to get back to Barnby Dun without getting lost again, much to my relief, and I felt very self-satisfied.

The small village of Kirk Bramwith is home to St Mary's church, an original Norman church dating from 1120. Its tower was constructed between the late thirteenth and early

Above left: St Cuthbert's church.

Above right: Snowboarding.

Below: The Don heads north from Fishlake.

Above: The Aire & Calder Canal heads towards Goole.

Right: The Dutch River looking east.

fourteenth century, and its bell was made in York in 1350. Robert Thompson, whose trademark mouse is internationally known, made most of the furniture, and twenty-seven mice can be found if one has time to search. A snowdrop festival is held here in mid-February each year, when hundreds of visitors arrive to see the carpet of snowdrops that covers the grounds. Although situated in Kirk Bramwith, St Mary's is the parish church of South Bramwith, Braithwaite, Fenwick and Moss.

STAINFORTH

From Bramwith, we travel eastwards to Stainforth, where the Stainforth & Keadby Canal and the River Done, Dun or Don, as it has variously been called in this area, come together again. The bridges that span these two waterways form one of the access roads to Fishlake, and several other villages within this lowland area. One of Vermuyden's earthworks is in evidence here.

As recorded in the previous chapter, Vermuyden's scheme to drain Hatfield Chase, an area of 180,000 acres, involved changing the direction of the River Don to the north to meet the River Aire, before reaching the River Ouse. However, this plan put Fishlake and several other lowland villages at great risk of flooding, and so the Dykesmarsh bank was constructed to contain the new river. It was built well to the east to allow for any overflow, but a flood in 1629 proved that this was insufficient; it became necessary to dig a new river from Newbridge to Goole to take the excess water. It took three years to construct what became known as the Dutch River, at a cost of £33,000. A great sluice was also inserted at the junction of the Don and Aire rivers to prevent tidal waters, which

used to flow to York along the Ouse, are being are diverted to the Don. Another sluice was built at the junction of the Dutch River with the River Ouse, in a further effort to control the tidal impact, but the sluice was swept away in another flood never to be replaced again.

The Dutch River eased some of the problems of the low-lying villages, but it wasn't until an even higher bank was built around 300 years later, by prisoners of war in the late 1940s, that villagers felt any sense of security. It is a rare event that threatens flooding these days, but it is still not outside the bounds of possibility, as the river is still tidal at Stainforth and several near misses have been recorded in recent years.

When Vermuyden's work began in 1626, a lake was situated between Stainforth and Thorne, which was fed by the River Don, but the system of drains and dykes installed allowed the Don to be diverted and this lake to be drained. Thorne, now situated on the Stainforth & Keadby Canal, is a centre for the building and conversion of barges and narrow boats, while Stainforth, once a thriving inland port, was turned over to coal mining and lost much of its business to Doncaster and Bawtry.

FISHLAKE

Crossing the bridges at Stainforth, the road runs alongside the river to Fishlake, though the river cannot be seen due to the high bank, which provides the highest ground in the area and serves as a floodplain. In 2007, this bank saved the village from flooding when the river overflowed, and was within half an inch of topping the bank as well. The picture, taken near Fishlake

The tidal Dutch River looking decidedly muddy as it approaches Goole.

The River Don reaches its journey's end as it flows into the Ouse.

Church in March 2008, demonstrates that the bank has uses beyond keeping the river out of the village.

Fishlake has some interesting features, and its name recalls a time when the river provided excellent fishing. St Cuthbert's church was built by the Normans, the doorway dating from 1171, and it is believed that the body of St Cuthbert rested here during the monks' retreat from Lindisfarne when the Danes invaded. Their wanderings lasted around seven years, and only ended when a final resting place was found for St Cuthbert in Durham Cathedral. Apart from the church, there are two old windmills, two medieval crosses, a Saxon pinfold and two pubs in the village.

Continuing on through the village, one arrives at the Jubilee Bridge, which recrosses the Don to join the A614, the canal having veered off towards Thorne by this time. Vermuyden used an old Roman channel, the Turnbridge Dyke, for this part of the river's new course. On a good day the river can be seen at either low or high tide, meandering lazily for miles across flat green countryside, Drax power station visible on the horizon.

Following the A614 for several miles, a sharp right-hand turn takes one alongside another high bank towards Rawcliffe Bridge,

Above: The Vermuyden Hotel.

Below: The swing bridge.

from where the Dutch River can be seen for the first time. Here there are two bridges and another stretch of water alongside it, which turned out to be the Aire & Calder Canal, with the M18 just visible in the distance to the west and Goole to the east.

It was low tide when I visited, and the dangers of venturing too close to the muddy banks suggested by 'Attercliffe Lad' in his blog in 2013 (www.riverdon.co.uk/writing/blog-on) could well be imagined.

The Dutch River is rarely used for navigation nowadays, the New Junction, Aire & Calder and the Stainforth & Keadby Canals being thought more suitable, and so it is mainly as a drain that it continues to be useful.

GOOLE

20 July 1826 was a very important day for Yorkshire. A new port, a new canal and a new town were all opened on the same day. Hundreds of people had come from Leeds, Hull and surrounding areas to celebrate. At half past three in the afternoon, fifty boats appeared coming along the canal from Ferrybridge. The first was flying a huge Union Jack and three banners, on which were written 'Success to the Port of Goole'. In the town, bands could be heard playing 'Rule Britannia' and guns were being fired in salute. According to one writer at the time, the visitor arriving by canal would see 'ships with their streamers flying, new streets stretching in various directions, mansions and shops of almost metropolitan appearance, immense warehouses with appendages for the loading and unloading of vessels...'

It is hard to imagine today how exciting the foundation of a new port at Goole must have been, around 200 years after Vermuyden's draining of Hatfield Chase and the building of the Dutch River. The above is a description provided by the Waterways Museum at Goole of what happened, and the excitement is palpable. You can almost see the crowds lining the bank and hear them shouting, 'Rule Britannia', as the boats sailed proudly along the canal with their banners flying.

I have already explained how the course of the River Don was changed in the seventeenth century, and how a sluice, preventing the River Ouse from entering the Dutch River, had been washed away in a flood never to be replaced. In fact, it was this seeming disaster that had enabled a port to be built here.

The textile merchants of West Yorkshire, who wanted to increase their trade by importing raw materials and exporting their finished goods, demanded a new and bigger port. They knew river transport would be more economical than road, but the rivers were shallow and navigation had to be improved if this method was to be made accessible to them.

Improvements to the River Aire and the River Calder had been made in the early 1700s, when sixteen locks were constructed, enabling the movement of goods in larger barges to Leeds and Wakefield respectively. These locks were enlarged as time went on, as was the depth of water, but they needed an outlet to the sea and Selby was increasingly lacking in both size and facilities. However, the opening of the Don into the Ouse at Goole provided just what they needed – deep water – which, along with a new canal, would service the docks that were now required to handle the booming trade.

A record of the many things being imported or exported all over the world between 1826 and 1899 includes iron, steel, vehicles, machinery, coal, coke, pitch, wool, textiles, fertilisers, chemicals, ores, minerals, timber and building materials. However, most of the trade today is with northern Europe.

By 1826, the new Aire & Calder Canal had been completed and Goole, along with the merchants, benefited. As locks increased in number and were made longer, six, ten, twelve or even twenty-one compartment boats carrying coal would be coupled together to be pushed or pulled by steam tugs, often interfering with other river traffic. In 1905, the network was completed as the New Junction Canal connected the Aire & Calder Canal with the River Don Navigation.

During this time, a new town was being planned by George Leather, who designed a square surrounded by grand buildings on three sides, and the dock on the fourth. Unfortunately, a lack of finance caused this plan to fail, with only one side, Aire Street, now the town centre, ever being completed. Shanty towns grew up to house those involved in building the docks, and within a short while it is estimated that 1,000 people were living in the twenty-three houses and sixty-one cottages that had been built.

Other industries soon grew up in the area and, as more and more people arrived to find work, additional accommodation was provided to cater for them and to reduce overcrowding. As Goole is almost at sea level, the land was built up to prevent flooding; while this alleviated some of the problems, some houses remained damp and unhealthy. Some unusual building features, which can still be seen today, demonstrate the problem, as houses were built on different levels; the fronts, higher than the back, looked quite impressive, while the backs remained liable to flooding. Rounded corners also prevented drays and carts from scraping the buildings as they negotiated corners, and there are still examples of these today.

The Vermuyden Hotel sits between the Aire & Calder Canal and the River Don near the Waterways Museum, and a swing bridge operates over the Don nearby.

4

VOLUNTEERING ACTIVITIES ON THE DON

Although the statutory agencies such as Yorkshire Water, the Environment Agency and British Waterways have played their part in improving the quality of water in the River Don, a great deal has also been achieved voluntarily by people who live and work alongside it. Where a need has been seen, groups have sprung up to improve their environment by clearing rubbish from footpaths and canals, laying hedges, pulling up invasive weeds, pruning overhanging branches from cycle tracks and maintaining public rights of way.

British Waterways, now The Canal and River Trust, and various wildlife trusts, actively encourage volunteer groups, providing supervision, training and support for funding applications for a whole range of environmental, social, health and leisure projects, some of which now provide permanent employment opportunities.

JOIN THE RIVERLUTION

Riverlution is an online forum run by the River Stewardship Company (RSC) for anyone interested in rivers and waterways around Sheffield, where the latest information can be found about volunteer days, corporate events, fun days, live music and other activities based around waterways in the area. News about improvements that are jointly taking place with local businesses can also be found.

The RSC came into being in 2008 with funding from the Environment Agency, the Key Fund and Natural England. One of its core aims was to secure private landowner investment, which it has since done with Sheffield Forgemasters, Meadowhall, British Land, Kelham Island Museum, Kelham Riverside (apartments), Brewery Wharf apartments and Sheffield City Council. As a result of this, it is now largely self-supporting, but grants are still required for new projects. The following are examples of their work.

When Sheffield Forgemasters International Ltd, based at Brightside Lane to the east of Sheffield, suffered significant flooding following the heavy rainfall in 2007, they worked with the Environment Agency to remove unsuitable vegetation and to plant more appropriate species. The RSC has now been contracted to carry out ongoing maintenance of this riverside, which will support wildlife and prevent debris from building up again. The area around the Meadowhall Shopping Centre also receives similar attention.

A grant was obtained from the Big Lottery Fund for The Blue Loop Community Project. This was a three-year project during which a self-sufficient group of Friends of the Blue Loop was formed, which is now supported by the Canal & River Trust. The Blue Loop is a circular walk taking in the River Don and the Sheffield & Tinsley Canal, between Blonk Street in Sheffield and the Tinsley Viaduct, where the canal runs parallel to the river before converging into it. The loop can be divided into three shorter walks if required, the Five Weirs Walk being part of this stretch of the river. As the waterways become ever cleaner, more and more species of wildlife are returning to it. At various times it is possible to see otters, herons, dragonflies, water fowl, kingfishers and butterflies, as well as the less desirable Japanese knotweed.

The RSC has organised numerous volunteer days when people help with the removal of invasive species, such as Himalayan balsam, as well as organising litter picks and the pruning of vegetation along the Five Weirs Walk.

Contact info@riverlution.org.uk for further information or sign up for a regular e-newsletter.

Volunteers on the Sheffield & Tinsley Canal, courtesy of the RSC.

SHEFFIELD INDUSTRIAL MUSEUMS TRUST

Volunteers are welcome to help out at Abbeydale Industrial Hamlet. Contact: ask@simt.co.uk for further information.

Work at Forgemasters and clearing debris below Norfolk Bridge, images courtesy of the RSC.

RIVER STEWARDSHIP COMPANY

Revitalising waterways

SWINTON LOCK ACTIVITY CENTRE

Information about the exciting work of this project preceded my visit on a hot summer's day in July, but the hospitality shown to me was beyond expectations. Based on the canal at Swinton, near Mexborough, I turned up unannounced and was immediately given basic information about their work. I was soon taken in hand by a volunteer named John, who was so enthusiastic and proud to show me around the site and the boats; I was even taken on a short boat journey along the canal, which enabled me to take photographs that otherwise would have been impossible.

John trains others, both young and old, in narrow-boat handling with a view to future volunteering. Now also responsible for health and safety and risk assessment, he and his colleagues use a narrow boat to offer six-week courses and taster days. But this is not all the activity centre offers. Short courses in many aspects of art and crafts, fishing, park maintenance, multimedia and photography are just a few of the subjects available at a reasonable price.

Ability is no bar to taking part in activities, and everyone is welcome on the boats, including wheelchair users. Young people from nine to eighteen years of age are all offered opportunities to look at life and learn from a different vantage point, and their community programme provides space for special events such as children's parties and boat trips.

Helped by the Heritage Lottery Fund, they have acquired a larger boat that can carry twelve passengers, and is fitted out with toilet and kitchen facilities.

I feel sure that none of this could happen without the input of many volunteers; the open days held each year provide a chance for anyone to visit and see the marvellous work they are doing (contact www.swintonlock.org.uk).

DENABY INGS NATURE RESERVE

This area, situated on Pastures Road, is the result of mining subsidence and is overseen by Yorkshire Wildlife Trust. It consists of a lake, reed beds, woodlands and a hay meadow, all of which attract many species of wildlife, including bitterns, avocets and spoonbills. Water levels can be manipulated to assist with flooding problems elsewhere. Orchids and other rare plants can be found here, and it is considered to be a naturalist's paradise. It is one of a series of nature reserves, the others being Old Moor, an RSPB site at Wath, Sprotbrough Flash in the Don Gorge, and Potteric Carr at Doncaster. Maintenance work is done by volunteers and anyone interested would be welcome to join (contact www.ywt.org.uk/reserves).

CONISBROUGH MILL PIECE CONSERVATION GROUP

Also known as EnTour, this group is concerned with environmental and tourism issues. The Mill Piece site used to house the 'Damhead Sawmill', which was in existence before 1838, and made bed poles, bobbins and other turned items using water power from Kearsley Brook, which runs into the River Don. It remained in the ownership of the Wilson family throughout most of its life, Charles Wilson of 'Willow Vale', Low Road, being the last owner.

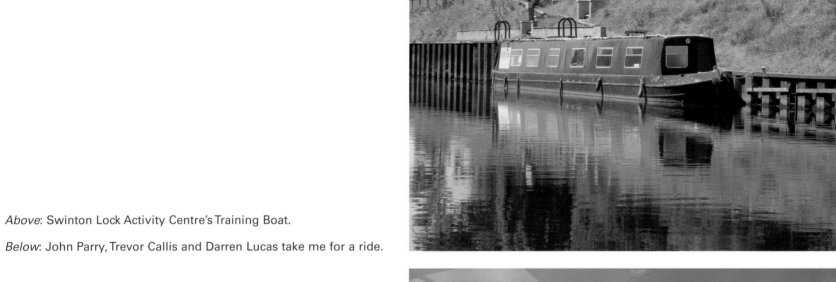

Above: Swinton Lock Activity Centre's Training Boat.

Below: John Parry, Trevor Callis and Darren Lucas take me for a ride.

Kearsley Brook.

The old mill was demolished a long time ago and the dam allowed to empty, but the brook continues to run into the river.

During the last century, this area was called the 'Dam Head'. Later it lost its 'head' and simply gave rise to an amusing true story.

A Mr Franklin had two allotments in which he grew vegetables and, in one nearest to the dam, flowers including roses.

One Sunday in the early 1900s, his granddaughter was seen carrying a bunch of these roses home when a lady, on her way to morning service, stopped to admire them. The little girl, with understandable pride, explained that they were from her grandfather's dam garden.

Such blasphemy would doubtless go unnoticed today, but ninety years ago things were very different and the child could not understand why she had been scolded.

In 2004, a group of Conisbrough people decided to get together to clear this small ponds that had become neglected and stagnant. Focusing on the need to remove litter and rubbish that had collected in the area, a partnership was formed with the local scout group and the Doncaster MBC Community First team; they filled a dozen skips. Reducing vegetation around the pond was the next step, but they discovered that the pond was contaminated with zinc and sulphur, and work had to stop until help to eliminate this could be found. In 2006, funding and planning permission were obtained, though this only covered two thirds of the cost and further fundraising was necessary before the work could be completed. The group celebrated its tenth anniversary in 2014 (contact info@entour.org.uk).

DON GORGE COMMUNITY GROUP

The river here is bordered on either side by steep limestone cliffs, which were carved out over thousands of years by the receding ice flows of the last Ice Age. For the purposes of this group, the Don Gorge is defined as being the stretch of the River Don from Conisbrough Viaduct in the west to the A1(M) motorway in the east. It is considered to be an area of outstanding natural beauty, and has several Sites of Special Scientific Interest (SSSI) within it.

This group began life as the Don Gorge Environmental Partnership (DGEP), with the aim of bringing together the many private owners and statutory bodies who had responsibilities in the area to encourage more co-operative working. After unsuccessfully trying to obtain a grant from the Heritage Lottery Fund to fund a development worker, it was eventually offered support by Doncaster Metropolitan Borough Council (DMBC). A worker was employed in June 2005, and a partnership was signed and sealed at the end of the year by British Waterways, the Environment Agency, DMBC, Doncaster Museums, Yorkshire Wildlife Trust, Lafarge Aggregates, Natural England and Don Gorge Community Group, known as the Don Gorge Strategic Partnership (DGSP). The DGEP was disbanded when the Don Gorge Community Group (DGCG) obtained charitable status.

For a variety of reasons, including the inauguration of a new, wider partnership known as Revival, the Strategic Partnership was itself disbanded in 2012 and the DGCG joined the new group. However, after a short period of time, funding for a co-ordinator could not be found and DGCG returned to their own resources. The main aim of the group is to promote, conserve and protect the Don Gorge through education and action, and to this end a group of conservation volunteers was set up who meet on a fortnightly basis. Originally supervised by a DMBC ranger, council cuts eventually made this untenable and they now have shared oversight from Yorkshire Wildlife Trust, the Canal & River Trust and DMBC.

However, the main thrust of the group has been related to tourism. The need for parking places has grown and the increasing popularity of fishing in the canal and river can make life especially difficult on match days. Several options have been pursued, but so

A happy band of volunteers.

far have come to nothing. The other need is for a visitor centre of some kind, but again nowhere has yet been found. Being a gorge brings its own problems of course, but the leisure opportunities are many: the Flash Nature Reserve with its bird hides and woodland walks are favourite visiting places, and the Wyre Lady pleasure boat and the Boat Inn both add to the pressures on the area. All these attractions are on the north side of the river and canal. While this is welcomed, the addition of a newly constructed fish and eel pass on the south side, in which members of the DGCG committee have been involved, will bring its own parking problems. Despite this, it is an area well worth a visit.

There are many walks within the woodlands, but a circular walk along the river bank, over Conisbrough Viaduct and back through Levitt Hagg, provides interest and good views over the river.

The Gorge is accessed via Boat Lane in Sprotborough village to the north bank and Mill Lane, Warmsworth, to the south bank. Road bridges across the canal and river link both banks, and allow access to the Trans Pennine Trail, which extends through the gorge on the northern bank of the river and canal (contact lizreeve@dongorgecommunitygroup.com).

THE FISHLAKE MONDAY CLUB

This self-supporting and self-motivating group have achieved a great deal on behalf of the village, from erecting gates on footpaths that were inaccessible unless one could climb a stile, planting 400 trees, erecting gravestones that had fallen or been pushed over, mowing the churchyard grass, laying a York stone floor to the rear of the parish church, restoring the pinfold and

planting wild flowers. It also receives donations from grateful residents who are helped with practical tasks, and this funds other work. Their latest project is to improve an area of the old River Don in the centre of the village, which was taken out when the river was straightened. Reinstating old drains that flow into the old riverbed will encourage the growth of reeds and hopefully deter nettles and bindweed, which are overrunning the area at present. The end result of this is that more birds, such as reed warblers, kingfishers, marsh harriers and bearded tits, will be encouraged, as well as damselflies, dragonflies and beetles. This work is supported by the Parish Council and members of the Monday Group also act as flood wardens (contact peter@trimingham.eu).

SOUTH YORKSHIRE TRANSPORT MUSEUM

A registered charity that I feel sure would welcome interested volunteers (contact www.sytm.co.uk).

THE WATERWAYS MUSEUM, GOOLE

The museum is based on the Aire & Calder Canal, and is well worth a visit. Many boats and barges are moored here, one of which can be toured.

Another boat of particular interest here is *Sobriety*, which is used by and for community groups. It is a Humber Keel, Sheffield Class, built in 1910 and measures 61 feet 6 inches by 15 feet 6 inches. The boat is a resource for personal development

The Monday Club start the clear up on the old river bed at Fishlake.

and the training of people from disadvantaged backgrounds. It has been particularly successful with those who have learning or mental health difficulties, young people at risk and those serving custodial sentences.

Volunteers are always welcome and contact can be made through their website (www.waterwaysmuseum.org.uk).

YORKSHIRE WILDLIFE TRUST AND THE CANAL & RIVER TRUST

These organisations are both very active along the riverside and, according to their particular purposes, have carried out numerous projects to improve water quality, wildlife habitats and leisure activities. YWT hosted Revival, the partnership set up to further co-operation between those interested in biodiversity and community groups working to improve their own specific areas, but unfortunately it could not be sustained. However, supervision is still provided by both Trusts for community volunteer groups, which enables work needing to be done to be co-ordinated.

YORKSHIRE WILDLIFE TRUST

Two projects of particular interest are Eels in Schools and the Repair of Riparian Habitat. In the former, tanks of eels were placed in schools until the eels were a suitable size to introduce to the river, and in the latter coir rolls were installed on the Ea Beck near Tilts. The rolls contained yellow flag iris, purple loosestrife, soft rush, lesser pond sedge, glyceria and reed canary grass, all of which would enrich the banks and improve the habitat for invertebrates and voles, as well as providing a refuge for young fish (contact www.ywt.org.uk).

Sobriety at the Waterways Museum, Goole.

Eels are looked after in schools before being transferred to a more permanent home.

Before and after; repairing the sides of the banks, photographs courtesy of Karen Eynon and Carys Hutton of Yorkshire Wildlife Trust.

THE CANAL & RIVER TRUST

The installation of fish and eel passes is a large part of the trust's work, and two have recently been completed at Meadowhall and Lower Sprotborough. Volunteers will continue to be involved in the maintenance of the site at Sprotborough, supervised by the Trust, when landscaping has been completed. (contact www.canalrivertrust.org.uk)

Right: Meadowhall Weir and fish pass.

Below: The fish and eel pass at Sprotborough Falls.

5

LEISURE ACTIVITIES ON THE DON

WALKING

The opportunities for leisure both on and alongside the River Don are many and varied. Although I have lived very near to it for most of my life, while researching this book I have been amazed to discover how little I actually knew about it. For this reason, I would recommend walking as the number one leisure pursuit. It is very easy to ride along in a car and see the river without really knowing or understanding any of its real nature or function. It is also very easy to look at a stretch of water and acknowledge its beauty or its foulness for a moment, and then dismiss it without much thought, but my journey has led me to places I had never been to before and enabled me to understand the different aspects and workings of it. To do this, it has been necessary to walk as well as ride. I feel sure that many ramblers, with whom I contrast 'strollers' like myself, have already covered much of what I have seen for the first time, but I would encourage others to search out the hidden places that are so fascinating and interesting.

CYCLING

This is the next best thing if one wants to reach inaccessible places, and the Trans Pennine Trail is ideal for this. It passes along many canals and rivers, including the River Don, as well as through historic towns and cities en route.

MESSING ABOUT IN BOATS

Water, of course, means messing about in boats, and this aspect of leisure is very well covered on the navigable stretches of the River Don. Whether by narrow boat, pleasure boat, canoe or rowing boat, quite a number of clubs exist to serve the needs of those who wish to participate.

Sheffield Amateur Rowing Club is based on Damflask Reservoir at Low Bradfield, near Sheffield, and its canoe club is at Oughtibridge, accessed from Forge Lane. Doncaster's rowing club is based at Hexthorpe.

The Strawberry Island Boat Club at Doncaster is a hive of activity for owners of narrow boats and other pleasure craft.

Above: A natural history walk.

Right: Cyclists on the Trans Pennine Trail at Lower Sprotborough.

A rowing competition and a fun day canoeing at Sprotborough.

The club house is inviting, and a friendly crowd shared their knowledge of boating with me, eventually leading me through the conservatory to the elbow of water that forms the island. Compared to 1969, when the site was derelict, it has now been developed to encompass moorings for a great many narrow boats and pleasure craft, all through the skills and maintenance of its members, who enjoy a good social life within its ambit. I was told that they could see you through from the cradle to the grave, with midwives and vicars and all skills in between represented within their membership. Having started on a barge, the club house then progressed to a wooden hut, which burnt down, and is now a fine building. The boat club obviously uses the waterways, but the island itself is full of allotments which have been available from long before the boats arrived.

The Wyre Lady takes trips, by arrangement, up and down the river from Lower Sprotborough in the Don Gorge. This photograph was taken on a Park and Float Day, when visitors to the gorge were encouraged to leave their cars at home.

Members enjoy a joke in the club house.

FISHING

This is one of the most popular leisure activities in the country, and Doncaster, Rotherham and Kilnhurst all have angling associations which fish the Don. Once too polluted to fish, the Don is now coming back to life, the presence of fish a good indication of this fact. Fish and eel passes are being constructed to enable salmon and other course fish to navigate the weirs. Upstream, 'small stream' trout and grayling are present, and stocking of the river with barbel began in 1990; around 5,000 have been introduced into the system by the Environment Agency. This shows a determination to regenerate the river, and the expectation is that barbel will gradually find their way to the upper reaches.

BIRD WATCHING

Yet another leisure pursuit that has many adherents who visit nature reserves – Denaby Ings and Sprotborough Flash being two – to observe and photograph the different birds, occasionally being surprised by an unexpected feathered visitor which sees twitchers rushing to the area.

SKETCHING

It is not unusual to see artists sketching or painting, but a day led by Sheila Bury brought a group of novices to the canal at Lower Sprotborough.

Don't forget that volunteering can also be a leisure pursuit. Many of those involved make good friends and look forward to their days out together, while doing something worthwhile. Christmas lunches are great too.

Above: Fishing at Kirk Bramwith, with the swing bridge in the distance.

Below left: A family of swans.

Opposite: Volunteers socialising at events held throughout the year.

6

THE FUTURE

DON CATCHMENT RIVERS TRUST: 'A NEW FUTURE FOR THE RIVER DON'

In 2011, ten pilot projects were formed to protect river wildlife and habitats by encouraging participation to improve the water environment: delivering a range of environmental benefits for the community; developing a shared understanding of the catchment priorities; and making sure participants feel a difference can be achieved. The Don Catchment Rivers Trust (DCRT) was one of these ten.

Having seen the need to develop such a scheme since 2004, the DCRT was formed in 2008, obtained charitable status in 2010 and was ready to begin. By December 2012, a plan had been delivered (www.dcrt.org.uk/the-don-network). The catchment-based approach is hosted by DCRT, but is in partnership with the Environment Agency and encompasses the rivers Don, Dearne and Rother, as well as many other streams and tributaries, such as the Loxley, Rivelin, Sheaf, Porter Brook, Dove and Went.

The aim of the trust is to restore the rivers to their former glory. As we have already discovered, during the Industrial Revolution the rivers were heavily engineered by straightening sections and building weirs to meet the needs of industry, leading to the pollution that gave the River Don the reputation of being the most polluted river in Europe. The ultimate wish would be that some weirs might be removed, and that the river might meander once again. In the meantime, the continuation of work to improve wildlife habitat by installing fish and eel passes, control of invasive species, removal of litter and habitat management all come within the trust's remit, which is committed to working with other agencies to achieve its ends.

A development grant has already been awarded from the Heritage Lottery Fund, with matching funding from the Environment Agency to the value of £102,000. However, a delivery grant of 84 per cent of the eventual cost of £1.4 million will still be needed if the desired work is to be completed, the remaining 16 per cent having to be raised from other sources.

The Living Heritage of the River Don, as the project has been named, has set three objectives: to encourage the return of salmon to the River Don, to formalise a long distance heritage trail from Doncaster to Sheffield, and to provide opportunities for people and communities to learn from the heritage that surrounds them along the river, helping them to develop new skills and interact with their heritage in new ways. I wish them well in their endeavours.

Membership of the Trust is open and free and those who join will receive invitations to meetings and be sent regular newsletters (contact info@dcrt.org.uk for further information).

ACKNOWLEDGEMENTS & REFERENCES

My grateful thanks go to everyone who has helped in the production of this book, but especially to Chris Firth OBE who very kindly checked my draft and ensured I hadn't made too many errors. Also to:

The Source: Holland, John, 'Tour of the Don' (Vol. 1, 1837).

Mexborough: J. R. Ashby of the Mexborough Heritage Society, and Giles H. Brearley of the Swinton & Kilnhurst Heritage Society.

Conisbrough: Gordon, Audrey and Kevin Mead; English Heritage for the Castle; and Conisbrough Mill Piece Conservation Group.

Sheffield: Sheffield Industrial Museums Trust and the River Stewardship Company.

Rotherham: Munford, Anthony P., *Rotherham – A Pictorial History*

Navigation: Taylor, Mike, *Memories of the Sheffield & South Yorkshire Navigation*, *Giant of the South Yorkshire Waterways* and *The Sheffield & South Yorkshire Navigation*

Don Gorge: The Don Gorge Community Group, Doncaster Naturalists' Society, *The Doncaster Naturalist: Don Gorge Special Edition*; Dennis Petty and Hebden Bridge Local History Society.

Doncaster: The Market Lavington Museum in Wiltshire.

The Old River Don: The LBBD Valence House Museum.

Goole: The Waterways Museum Trust.

Leisure: The Strawberry Island Boat Club.

The following agencies have also added to and enhanced my general knowledge of the river during the past ten to fifteen years: the Canal & River Trust, the Don Catchment Rivers Trust, the Doncaster Metropolitan Borough Council, the Environment Agency, Pennine Waterways and their website, the River Stewardship Company and the Yorkshire Wildlife Trust.

Lastly, but by no means least, thank you to all the individuals I met along the way, who provided snippets of information that I wouldn't otherwise have heard.